THIS MORTAL COIL

THIS MORTAL COIL

THE HUMAN BODY IN HISTORY AND CULTURE

FAY BOUND ALBERTI

OXFORD UNIVERSITY PRESS

OXFORD
UNIVERSITY PRESS

Oxford University Press is a department of the University of Oxford. It furthers the University's objective of excellence in research, scholarship, and education by publishing worldwide. Oxford is a registered trade mark of Oxford University Press in the UK and certain other countries.

Published in the United States of America by Oxford University Press
198 Madison Avenue, New York, NY 10016, United States of America.

Library of Congress Cataloging-in-Publication Data
Names: Alberti, Fay Bound, 1971- , author.
Title: This mortal coil : the human body in history and culture /
Fay Bound Alberti.
Description: New York : Oxford University Press, [2016] | Includes
bibliographical references and index.
Identifiers: LCCN 2015042473 | ISBN 9780199793396 (hardback : alk. paper)
Subjects: | MESH: Anatomy—history. | Human Body. | Health Knowledge,
Attitudes, Practice. | History, 20th Century. | Physiological Phenomena.
Classification: LCC QM23.2 | NLM QS 11.1 | DDC 611—dc23
LC record available at http://lccn.loc.gov/2015042473

1 3 5 7 9 8 6 4 2
Printed by Edward's Brothers, USA

For Rosalie Farnell-Leonard, my 'glamorous grandmother'
3 April 1921–27 July 2014

ACKNOWLEDGEMENTS

Research for this book was facilitated by a Wellcome Trust project grant. During my Senior Research Fellowship at Queen Mary University of London I co-founded the Centre for the History of Emotions, the first interdisciplinary emotions centre in the United Kindom. The Queen Mary History Department and the Emotions centre are home to many creative and generous scholars. To Thomas Dixon, Rhodri Hayward, Colin Jones, Miri Rubin, and Tiffany Watt-Smith, I owe sincere thanks and friendship.

I am grateful to Luciana O'Flaherty for inviting me to write this book, for her patience when its completion was delayed by life, and for her critical comments. Thanks also to her colleagues at Oxford University Press, especially Matthew Cotton and Erica Martin. I would like to thank my anonymous reviewers wholeheartedly, and Miranda Bethell for her careful copy-editing.

Thank you to the staff at the Royal College of Surgeons of England, especially Louise King, Ruth Neave, and Brian Morgan for their kind assistance on the history of cosmetic surgery. Ian Burkitt and Fernando Vidal were generous in their time and ideas in talking through their work, for which I am grateful. Many thanks to Mark Jackson for his comments on 'Beauty and the Breast', a version of which will appear in his forthcoming edited collection on the history of disease.

On a personal note, I am grateful to the following people for their support, suggestions and a well-placed word at just the right time: Karen Alberti, Emma Alberti, George Alberti, Jo Alberti, Stephanie Amiel, Louise Anderson, Nikki Bandey, Joanna Bourke, Jenny Calcoen, David Clayton, Lauren Couch, Lol Crawley, Lesley Dean, Laura Gowing, Jeanette Gregory, Javier Moscoso, Paddy Ricard, Matthew Shaw, Wiebke Thormahlen, Sandra Vigon, and Paul Woodgate.

In 2014 my daughter went through a life-changing surgical procedure that I have written about in this book. During that time the dedicated staff at Great Ormond Street Hospital made everything more bearable, especially Rory Philbin and the nursing team. Stewart Tucker is a highly skilled surgeon and I thank him and his colleagues for getting it right.

Putting the final touches to a manuscript can be gruelling, and I thank Sam Alberti for his support, especially in the concluding stages. I appreciate his bibliographic and proofreading help, his useful commentaries and his keen eye for detail.

To my beautiful children, Millie Bound and Jacob George Alberti: mind, body, soul, whatever there is, whatever I have: it's yours.

CONTENTS

LIST OF ILLUSTRATIONS

Introduction

The Body in Parts

To be, or not to be: that is the question:
Whether 'tis nobler in the mind to suffer
The slings and arrows of outrageous fortune,
Or to take arms against a sea of troubles,
And by opposing end them? To dic: to slccp;
No more; and by a sleep to say we end
The heartache and the thousand natural shocks
That flesh is heir to, 'tis a consummation
Devoutly to be wish'd. To die, to sleep;
To sleep: perchance to dream: ay, there's the rub;
For in that sleep of death what dreams may come,
When we have shuffled off this mortal coil,
Must give us pause: there's the respect
That makes calamity of so long life[1]

The playwright William Shakespeare is often depicted as the first 'modern' writer, with *Hamlet* as a particular example of anti-mediaeval modernity.[2] The above soliloquy is one of the most famous and the most debated in literary history. It is supposed to mark Hamlet out as Everyman (or woman), suggesting that we all face existential dilemmas about how to act and engage in an uncertain world.[3] Hamlet's fears and anxieties are our own, most notably for historians who claim passions like fear and anger are the same across times and cultures.[4] According to this logic, our bodily experiences must also be unchanged. Like Hamlet we each inhabit a physical body, a mass of flesh, bones and blood that takes us through this 'mortal coil' that is the 'bustle and turmoil' of everyday life.[5] Like Hamlet too, we are tied to the world by our bodies and by all the concerns that 'flesh is heir to'. Our bodies belong to us, in life at least. After death some of us give them up, or parts of them, donating our organs to help others live. Although many people are excluded from blood, tissue, or organ donation because of health and cultural or religious beliefs, there is a widespread presumption that we are all potential donors of spare parts.[6]

Few of us are happy with our bodies, despite their endless, dedicated service.[7] We want them to weigh less or to be more slender or muscle-bound. We want our faces to be more beautiful, our jaws to be squarer and our breasts bigger; we try to remove our wrinkles, veins, stretch marks, spots, and scars. For some, the search for the 'perfect body' is a quest that begins in the gym and ends in the operating theatre. We might even dream of escaping our bodies entirely, and perhaps some do, claiming to operate on a more spiritual plane without hunger, anxiety, or even pain.[8] But for most of us our bodies are the inescapable material reality that we live with and in. Whether regarded as 'machines' or 'temples', they get us from A to B and from birth to death. They also give us considerable pleasure along the way, in kissing our lovers, hugging our friends; in eating and drinking; and in endless forms of physical exercise and expression, from dancing and singing to running, pole-vaulting, and sex.

We know that conceptions of the perfect body are cultural: perceptions of beauty and health vary across time and continents. But it is taken for granted that humans through history have felt relatively similarly about their bodies, and what they can experience. After all, why should we feel any differently from our forebears? Like them, our hearts beat within our chests, our cheeks blush, our tongues taste, our stomachs digest and our lusts are inflamed. Some people claim to have penises with 'minds of their own'; others possess vaginas that are by and large described by their absence (i.e. the lack of a penis).[9] Like the similarly problematic concept of 'human nature', the body is an unchanging entity in a changing world. For many—certainly outside academic debates about bodies and emotions—the idea that 'the body' has a history is faintly absurd. Bodies just *are*: stable and solid vehicles for our emotions, our memories, and even our souls, if such things exist.

Yet the idea that Shakespeare is our 'contemporary', as was claimed in the 1960s, ignores the considerable differences in ideology, religion, medicine, politics, philosophy, and art that have taken place between the seventeenth and the twenty-first centuries.[10] As I will show in this book, Shakespearean views of the body, mind, and soul were rather different from our own, though they might survive in the language of the passions, say, and the metaphors of mind and body. The history of the body has become a popular subject for academics, especially cultural historians, as have debates about the significance of clothing, gesture, dance, tattooing, and piercing.[11] Most of these subjects can be regarded as things that we do with the body; the ways we move in and through society. But some have to do with the stuff of the body itself, such as emotions, the senses, pain, insanity, cardiac failure, the bowels, and blood.[12]

A particular influence on the academic study of the body in the twenty-first century is the emergence of new technologies.[13] Scientific advances in medicine and genetics including cryonics and cloning, and nanotechnology and transplantation technologies, as well as health and ageing crises, present ethical questions about our bodies and the limitations of the human. These critiques build on late twentieth-century challenges to the positioning of women, ethnic minorities, workers, people with disabilities, homosexuals, transgender people, and others who feel their bodies do not fit into traditional scientific discourses.[14] There is also much writing about the body as a product of language. This claims there is nothing 'real' about the body other than the stories that we tell.[15]

This Mortal Coil is situated within the histories of medicine, pathology, and the body as well as the histories of emotions and culture. It uses not only medical scholarship but also literary evidence to understand how and why our beliefs about the body have emerged and how those beliefs impact on our lived experiences. It argues that the ways we inhabit the body have changed, along with the meanings bestowed on certain organs like the heart, the brain, the spine, and even the tongue. Throughout history the body, with all its nerves, its veins and arteries, its organs and pathways, its fears and worries, is terrain that has been mapped in different ways, while the analogy of the body as foreign land to be explored and understood reinforces the idea of the anatomist as explorer and civilizer.[16] For centuries we believed that we possessed souls that were part of the body and inseparable from it. Now we exist in our heads, and our bodies are vessels for that uncertain and elusive thing we call our 'selves', at least in the West.

This book is explicitly situated in Britain, dealing in part with North America. It explores how we view our bodies as subjective or metaphysical entities; we might worry about the size of our breasts and whether we look old or fat, but many of us also have a sense of ourselves above and beyond our physicality. This immaterial self is usually present in the struggle between the eternal mind (translated as reason) and the mechanical body (that includes the passions): a struggle first identified by the French philosopher René Descartes fifty years after *Hamlet* was first published. What has become known as Cartesianism divided the world into three states of existence: that inhabited by the physical body (matter, which includes the operation of the passions), that inhabited by the mind (which includes reason), and that inhabited by God. It is the first two of these that concern me here: the separation of the realms of body and mind. This distinction is untenable for many reasons, not least because reason and emotion are not entirely opposed states. Without emotion, as one leading neuroscientist has put it, reason 'turns out to be even more flawed than when emotion plays bad tricks on our decisions'.[17]

And yet our entire Western medical system is based on the separation of mind and body. Since the nineteenth century the history of modern medicine has been one of narrowing focus and specialty. Hundreds of years ago doctors treated the whole person, seeing disease as due to disruption in the individual body. The rise of scientific medicine made internal threats into external ones; diseases became physical invaders located in specific organs, tissues, and cells. With the notable exception of general practitioners, doctors—by which principally I mean hospital physicians—became specialists working on specific body parts or disease classes.[18] The benefits of biomedicine are undeniable, and yet its weaknesses are built into its very structure. Today's governmental funding, research, building, and personnel resources form separate systems to deal with mental health (psychiatry and psychology), the heart (cardiology), the brain (neuroscience), the skin (dermatology), the gut (gastroenterology), and so on. While the so-called 'clinical encounter', an interaction between patient and specialist, has become the lynchpin of modern medicine, there is often something of a 'disconnect' in diagnosis and treatment.[19] And in the busy hospital setting there is evidence that errors are made by specialists trying to 'pull' cases towards their speciality, overlooking symptoms that seem irrelevant.[20] The patient, who might experience dis-ease as a whole body–mind affliction, can inevitably feel marginalized.

This is not just a book about historical scientific pathology, then, though its influence is felt in each chapter. I am interested in how the body and its parts have become subject to the classifying gaze of *difference*, which originated in scientific discourse but which now covers social forms of pathology.[21] Previously ignored characteristics, from having small breasts to being overweight, have become recognizable medical and social 'conditions'. In the field of medicine this diagnosis of the abnormal took place through case histories that made the individual a cypher for the universal. One example in this book is that of Charles Uncle, a young boy with skin sores who became a nineteenth-century representative of a specific disease classification: tuberculosis elephantiasis. His case is considered in Chapter 6.

Becoming a patient was fraught with difficulties, emotional, practical, and logistic, in the past, as in the present. This is perhaps even more jarring for those working in the medical profession. One hospital physician, Charlotte Yeh, has detailed her experience on the other side of the doctor/patient divide. After a traumatic accident she found she was no longer treated as an individual person, but subjected to 'piecemeal evaluation': 'I was a participant-observer in emergency care, with a big-picture window into how well our health care system does or doesn't work', she writes. 'There's just something about being boarded on a gurney in a hospital hallway for fifteen hours that gets one thinking about paradigm shifts'.[22] The experience is no better for those of us without medical training and unaware of what protocols should be being followed.

This book begins with a case study that is particularly familiar to me. In 2014 my fifteen-year-old daughter Millie, diagnosed with severe scoliosis at the age of twelve, was admitted to hospital for a 'spinal fusion with instrumentation'. Rather than growing straight her spine had curved and twisted, giving her back pain, knee pain, and poor blood circulation. Worst of all from the perspective of a teenage girl, one shoulder blade stuck out as a result of her ribcage rotating. Hemlines hung poorly and dresses could not have zips: their straightness highlighted her asymmetry. Spinal fusion was advised to fix all of these problems by mechanically straightening her spine, pinning it into place with metal rods and screws. Full recovery would take two years and the procedure itself was full of identified risks, but the predicted outcome of not having surgery was worse: increased spinal curvature as she grew older, increased pain, delays in healing, and potential complications in the event of pregnancy.

The likelihood of each of these physical risks was quantifiable, as were the surgical risks, including infection, loss of feeling, autoimmune rejection of the metal 'hardware', and paralysis. But there was no accounting system for the tears of my daughter as she sat passively in the surgeon's office, listening to such depersonalized terms as 'deformity' and 'lung drain'. Surgeons have not historically been trained in communication skills: they can engage with the body, but often not with the person to whom it belongs.[23] The treatment of my daughter's condition exclusively in orthopaedics, moreover, meant that any potential 'non-spine' issues (neuralgic, pulmonary, cardiac, urinary, renal or psychological) would require referral to another specialist.[24] Yet there is evidence that scoliosis is linked to depression, anxiety, and poor body image, especially in the case of adolescents.[25]

Scoliosis is a condition associated most commonly with Shakespeare's Richard III: the so-called hunchback king. Chapter 1 traces the history of scoliosis back to the ancient Greeks and the writings of Hippocrates and Galen. It suggests that although the meanings of bones and the spine have changed, the 'skeleton-as-scaffolding' metaphor has not. Nor have the therapeutics used in scoliosis treatment. In the ancient world as today the emphasis was on mechanically straightening the spine, using a ladder, a brace, or an operating table. What is new is surgical intervention, which like all surgery was dependent on the development of antisepsis and anaesthetics as well as skills and training.[26] The Royal National Orthopaedic Hospital traces its history back to 1838, but early orthopaedics was essentially 'bone-setting', a practice by which barber-surgeons manually fixed and reset fractures. It was not until the outbreak of the First World War, when a mass of casualties needed urgent treatment, that orthopaedic surgery in the modern sense was established.[27]

The same is true of other specialties. Chapter 2 shows how the earliest forms of plastic surgery rehabilitated the faces and limbs of soldiers wounded in battle. The skills acquired by the end of the Second World War found a market in consumers seeking a reprieve from another of life's trials: ageing. One of the ways disciplines were established was by determining what normal and abnormal or healthy and diseased should look like.

The abnormal and the pathological were determined by comparison—of case studies of living patients as well as in the dissection of the dead.[28] The subsequent emergence of cosmetic, rather than plastic surgery, was a result of previously normal conditions being identified as deviations from the norm. Having small breasts acquired a name—hypomastia—and women who 'suffered' from the complaint were also often diagnosed with mental illness, a correlation that cries out for further research. With two major health scares linked to silicone implants in Britain and North America, and a continued lack of governmental regulation of cosmetic surgery in addition to complex issues around consumer choice and responsibility, women's bodies are at the centre of a series of ethical crises. And not for the first time.

The arguments presented in this book have a feminist perspective; I am particularly interested in the construction of the female body. I do not, for instance, examine the cultural history of the penis: the language of sex remains phallocentric and 'thrusting manhood' is everywhere.[29] I have chosen to focus mostly on female bodies because many historical medical, and scientific ideas about female bodies stay with us, naturalizing inequalities between the sexes. This is seen in Chapter 3, an anatomical and cultural history of women's genitals: the vagina, the vulva, the hymen, and the clitoris. In Shakespeare's writings the vagina is a 'deep pit' or as a 'nothing', reflecting contemporary anxieties about female sexuality. Women's sexual voraciousness was a symbolic castration to men—made physical in the early twentieth century by the psychoanalyst Sigmund Freud's reference to the *vagina dentate* or the toothed vagina. Female sexual pleasure is problematic; female genitalia contested. The very existence of the hymen and the clitoris has been historically contentious. The language used to describe female sexual function remains opaque, and anatomical teaching is overlaid with social and political ideals about women's roles as mothers, wives, and reproductive units. This 'othering' of the female form is most strikingly apparent in the porn industry, where women's bodies are presented as a series of waxed, plasticized, and oddly disembodied orifices. How troubling, then, that there are such close parallels between elective surgery to 'tidy up' women's genitals—most notably labiaplasty, the reduction of the internal labia in pursuit of a 'Barbie' aesthetic—and the global incidence of female genital mutilation.

Several chapters in this book touch on surgical interventions. Like orthopaedic and plastic surgery, cardiac surgery was a product of the early twentieth century. Today new technologies explore how robots might remove human error in surgery, enabling maximum precision with minimal incision. One example is the Polish cardio-robot 'Robin Heart'.[30] In the 1950s and 1960s the global race was on to perform the first successful human heart transplant. The winner was the South African surgeon Christiaan Neethling Barnard.[31] In principle this should have been an uncontroversial, if surgically complex, procedure. After all, twentieth-century anatomy viewed the heart as a pump, responsible for moving the blood around the body. But the heart was and is far more than that, at least in the popular imagination.

In Chapter 4 I explore how the heart's shape, physiology, and iconography draws on questions of the soul and the personality as much as its bloody physicality. The fear of black people becoming 'spare parts' for whites is not prevalent today, as it was in apartheid South Africa. But heart transplants still carry the suspicion that more than the organ is being transplanted: perhaps even the personality, memories, and feelings of the donor. We cannot understand the heart's significance, or the modern-day language of emotion, without reference to the organ's complex history. Since the classical period a clear philosophical and theological tradition has held the heart, not the brain, as the organ most associated with our emotions, our selves, and even our souls. Hearts could be 'hard' or 'soft', 'cold' or 'warm', like their owners; part of an holistic medical tradition that regarded the body and the mind within an interactive system of fluids and humours.

The humoral model of emotion physiology remained intact for nearly two thousand years between the second and the nineteenth centuries. It is integral to many of the following chapters, so warrants some elaboration here. Humoral medicine was codified from the work of Hippocrates by Aelius Galenus or Claudius Galenus; better known as Galen of Pergamon, a prominent Greek physician, surgeon, and philosopher who lived in the Roman Empire. Galenic principles, doctrines, and concepts about the body and its functions dominated medical theory throughout Europe until at least the early nineteenth century.[32] For Galen the body was a 'little world' or microcosm of the universe, the tripartite divisions of heaven, sky, and earth corresponding to the three main parts of the human body: the head (reason), the breast (heart) and the lower body (nourishment and procreation).[33] The body had all the qualities that made up the 'greater world' of fire, air, water, and earth.[34] Four qualities—hot, cold, moist, and dry—inhered within these elements: fire was hot and dry; air was hot and moist; water was cold and moist; and earth was cold and dry. In each individual, these characteristics received the form of 'humours', which coursed through the body: blood, which was hot and moist like air; choler (or yellow bile), which was hot and dry like fire; phlegm, which was cold and moist like water; and melancholy (or black bile), which was cold and dry like earth.[35]

The proportional balance of these humours within each individual was partly innate and partly a result of nurture or environment; a product of heredity, age, sex, and what contemporaries called the six 'non-naturals': air; food and drink; exercise and rest; sleep and waking; evacuation and repletion; and passions of the soul. Although the passions acted on the spirits and humours they were also influenced by, and a product of, humoral balance. And an individual's humoral balance was partly environmental. As the seventeenth-century English philosopher Thomas Hobbes put it, their proportions 'proceedeth partly from the different constitution of the body, and partly from different Education'.[36] Nevertheless, the individual composition of humours determined a person's psychological and emotional state, humours being produced in the liver and coursing through the veins, mingling with the blood and affecting the mind, soul, and the body. A disproportionate amount of any one of the humours led to constitutional imbalance, including illness, as well as extreme emotional states. Humours and emotions were inseparably linked. In the words of the English writer Thomas Wright, 'passions ingender [*sic*] humors and humours bred [breed] passions'.[37]

Although the language of personality types was not commonly discussed until the eighteenth century, early modern men and women were characterized in ways that remain familiar: a high level of yellow bile made men and women subject to anger (choleric) and black bile to sadness (melancholic), while an excess of blood or phlegm made one sanguine (and prone to love-sickness) or phlegmatic. Each age and sex had 'prevailing humours' as 'the manners of the soul follow the temperature of the body'.[38] This was why young men were described as 'hot, incontinent and bold, old men are cold, covetous and cautious, women are envious, proud and inconstant—and these differences rest on differences in corporeal makeup'.[39]

Humoralism was inherently gendered. There were specific emotional tendencies associated with each sex, and those tendencies were apparent in physical differences. Women tended towards a phlegmatic, or cold and moist disposition, since their bodies were fleshier, softer and weaker than those of men, their hair longer, their faces paler, and their skin more moist. The greater passivity of women also made them more subject to such emotional extremes as hysteria.[40] Men, by contrast, with their leaner bodies and drier complexions, tended to display qualities of courage and anger. People were also prone to varying emotional behaviours during the course of their lives. This was because their natural heat diminished over time. As Wright put it, 'younge men generally are arrogant, prowde, prodigall, incontinent', and old men 'subject to sadnesse caused by their coldness of blood'.[41] Emotional expressions were also skewed by age and gender. The preponderance of water in women's constitution (especially young women) meant that they were more prone to tears, and also to sudden, irrational rages since women's flesh 'is loose, soft and tender, so that the choler being kindled, presently speeds all the body over, and causeth a sudden boyling of the blood about the heart'.[42] Women's anger soon passed, however, since (like old men), they lacked the heat to sustain the emotion.

There are obvious parallels between the 'heartache and the thousand natural shocks' of Hamlet's body and our own, in the association of women with tears and men with anger, for instance, as well as the languages of the passions. Yet the differences between the humoral body and our own can be seen in the representation of mental illness. In *Hamlet* Ophelia's madness is dramatized in physiological terms. It comes from 'the poison of deep grief', according to King Claudius: the body being overwhelmed like an off-guard army as 'when sorrows come, they come not single spies, but in battalions'.[43] Ophelia is 'divided from herself and her fair judgement' and cast like a 'mere beast' into the world of the flesh. She is all 'winks and nods and gestures', beating her heart over lost love a familiar and gendered sign of dismay and female abandonment. Drowning was the number one cause of suicidal death among early modern women.[44] Symbolically Ophelia's watery end was a reminder of the natural wetness of women's constitutions: lacking men's heat they were far more likely to succumb to depression. Melancholic humours were accumulated in the body, summoned by the heart as a response to the soul's grief, concocted in the liver, and sent to the brain where they overcame reason.[45]

Shakespeare's imagery might have resonance today, but his vision of emotions as products of the body are at odds with modern explanations. From the late nineteenth century, emotional and intellectual experiences were not linked to the physical body but to the brain, and even to specific locations within the brain. A graphic example discussed in Chapter 5 was the case of the railway worker Phineas P. Gage.[46] An industrial accident in

1848 nearly cost Gage his life after a metal bar shot through his head; none of the witnesses or doctors expected him to survive. He lived for another twelve years, but the accident cost Gage his ability to regulate his behaviour and his emotions. As a result of those perceived failings he also lost his job. During Gage's lifetime no credence was given to the idea that his changed behaviour was linked to frontal lobe damage. Only many years after his death was this association made explicit and the material brain associated with personalities, emotions, and even selfhood.

When I use the term 'material' here and throughout this book, I am referring to the physical structure of the body as opposed to its spiritual and immaterial aspects. However, I am not suggesting that the material body was always secular; for some theorists (the vitalists of the eighteenth century being a case in point), a sacred spirit was entirely compatible with its physical form. Moreover, even those (like the mechanists) who did not believe the body had a soul viewed the human body as a sacred vessel and an illustration of the work of the creator, a theme taken up in Chapter 5, below.

The arrival of the brain-centred self might seem a fait accompli. But the mind is not always reducible to brain, as seen in my discussion of earlier theories. Debates over the role of the immaterial mind continue. Moreover, the idea of localization was not entirely a product of the nineteenth century. It existed centuries before, albeit in different language. And Gage's case might have made an impact earlier, had he not been buried without autopsy. His skull was exhumed by his attending surgeon, John Martyn Harlow, who had received some training in phrenology, a pseudoscience that used elaborate measurements of the skull to associate parts of the brain with behaviour and aptitude.[47] Craniometry and craniology, too, focused on intellectual hierarchies as detected through the skull's physical characteristics. In each case scientific classification of the skull and brain justified sexism and racism, with the cranial structure of women and non-Europeans found to be inferior to their white, European male counterparts.

Gender and race were manifest in many of the body's organs. In the early modern period those differences were rooted in humoral balance, whereas by the nineteenth century they were found in the physical structure of the body itself. This can be crudely sketched as a transition from fluids to fibres, humours to nerves and heart to brain.[48] Psychological and physical, and even moral associations were etched on the body, including concepts of evil and good as related to colour that were well established by Shakespeare's time. In *Othello*, blackness was associated with sexual power, passion, and violence, a stark contrast to the white, virginal skin of Queen Elizabeth I.[49] Early racial traits were related to humoral and climate differences that in the nineteenth century were overlaid by the scientific concept of race.[50] As Chapter 6 shows, this redefinition of race was part of a broader rethinking of the anatomy and pathology of the skin to signify difference as well as sensory perception.

Today the skin is commonly described as the body's most extensive organ. It is not only the psychological and material boundary of our bodies, but also protects us physically, provides sensory perception, communicates emotions, and acts as an environmental filter.[51] What a move away from ancient theory and from the view of Aristotle, for whom skin was a hardening and drying of the external body, akin to the film that forms on a bowl of porridge. The skin has become one of the most significant material indicators of ethnicity, gender, age, experience, and social identity. Women's skin is traditionally expected to be soft, smooth, and hairless and men's to be rugged and hairy, though the latter is changing with a growing skin-care market targeting male grooming.[52] In both sexes, lighter skin tones are touted as the ideal, whether in the world of cosmetic advertising or in education and employment prospects.[53] At the level of experience, the 'fairer sex' is far more than metaphor.[54] White skin is associated with civilization and purity while dark skin represents animalism and brutality.[55] This racial stereotyping of black-skinned bodies with darkness and base sexuality was evident in the scientific racism of the nineteenth century, and in the appalling treatment of African women like Saartje Baartman, the so-called 'Hottentot Venus', a woman famously reduced to her enlarged genitalia and pronounced buttocks.[56]

We do not normally think of the tongue as being laden with the same gender, class, and age-based meanings as the rest of the body. Yet Chapter 7 considers some of these meanings in an episodic history of the tongue as a social and political weapon, a conveyer of taste, and a measure of health and disease between Shakespeare's time and our own. Our own tongues change during our lives, and not only because they weaken with age.[57] What we eat affects our taste buds, too much sugar and salt dulling them to more nuanced flavours.[58] The idea that physiological changes, in this case in taste, result from environmental factors is a significant one. It might even influence our views on obesity, greed, and the gut.

Chapter 8 explores the complex, mutually interrelated histories of fatness and the gut. In an age of 'globesity', in which a high percentage of people are obese, attitudes towards fatness seem straightforward: obesity is a direct product of eating too much (particularly of the wrong food) and exercising too little.[59] The stigmatization of fat people is part of a culture of blame linked to the moralizing theories of will power and greed. This perspective is largely a post-industrial one: we are locked into a system of efficiency metaphors in which fatness equals waste in an age of production. The language of obesity is not only associated with laziness or ignorance, but also by presumptions about gender, race, and class.[60] The thermodynamic, calorie-based model is based on viewing the body as a machine. It has little regard for the psychological causes of obesity or the complex biochemistry of the body and its digestive processes.

Today, research suggests that obesity is not simply about balancing what we eat with how much we move. Like the heart, the gut has been given a 'second brain'. The stomach and the intestines are seen to do far more than process food, absorb nutrients and expel waste. They are part of the entire gastrointestinal system, governed by the enteric nervous system (ENS), which has up to 600 million neurons, the same number as the spinal cord. The gut has become a mass communication centre, signalling to the brain, responsible for the immune system as well as the maintenance of the hormones (which are arguably our modern-day humours). 'Gut feelings' have been given material basis, as has the belief that food can cause depression—not in the same way as Shakespeare's time, by sending noxious humours to the brain, but by altering our biochemistry. It is not so much what we eat and how we move, in this narrative, but the nerves of the gut, its chemical balance and even its intestinal flora that causes obesity. The gut, the neglected middle child of the feeling trinity—the head, the heart, and the belly—is coming into its own.

The chapters thereby take us on a narrative journey from inside out, from our very core to the surface of our body and the boundaries between self and other. Beginning at the spine and its perceived materiality as the foundation of our physical selves, we move to the breasts and vagina, both of which are heavily invested with cultural meanings as sexual and procreative organs. The heart and brain are considered next as parallel and often competing objects in the construction of personal identity. Traditionally viewed as opposites, heart and brain, emotions and reason, have more in common with one another than we might imagine. And from the origins of the self we move to skin and tongue; from the boundary of our physical and material self to the medium of expression by which many, though not all of us, communicate with the social world. Ending with fat and the gut, as this book does, might seem a surprising way to conclude, and yet there is an important rationale. Fatness in the modern West is vilified. It is a visible manifestation of a lack of control, of an inability to contain one's physical self within appropriate boundaries. The gut is often symbolic of that lack of boundaries, associated as it is with gluttony and decadence, with over-eating and inefficiency. And yet there are important links between fatness and the gut that bring us back to the theme of holism: fat is more than energy in versus energy out (which is itself a recent historical concept), and our guts are more than factories for the consumption of food. The emergence of the 'gut–brain' brings together our emotions and our rational appetites, our psychological and physical experiences in profound ways—and not only through 'gut-wrenching' fear or 'butterflies' in our stomachs. It provides a model by which the body can be seen to impact on the brain, as well as the brain on the rest of the body, and a narrative framework by which the language of our *somas* can become a legitimate part of our knowledge about the world.

A number of important strands run between each of the following chapters. These include differentiating between male and female bodies: from skeletons to brains, from skin to genitalia, male and female bodies are regarded as absolutely different. These divisions not only reinforce assumptions about men's and women's 'natural' capabilities; they also have devastating effects on the unknown numbers of individuals that identify as transgender or intersex, or simply don't relate to a binary model of gender difference.[61] The process by which some body parts and experiences are made pathological is another important and recurring theme, and these discourses are not only derived from medicine and science but also from visual and material culture, philosophy, religion, and consumer capitalism. Thus cosmetic breast augmentation in the 1950s combined aesthetic judgements about the female breast as a site of nurture and sexual pleasure with the possibility of perfection. This goal became possible for the first time by surgical prowess and technologies like silicone, and the preparedness—or perceived entitlement—of women to become enlightened consumers in the medical marketplace.[62]

One of the questions most dreaded by historians is 'so what?'[63] Why does it matter if the body and its parts are layered with meanings, or if there has been contention throughout history over the primacy of the body's organs and the relationship between the material and the immaterial realms? Why should we care if we separate mind and body in ways that make the body a machine or the vagina a passive recipient of the penis?

I believe that it does matter, and we should care. The dominant language of sex that eliminates female pleasure from reproduction, for instance, not only denies the breadth of women's experiences but also perpetuates misogynistic ideas about women's sexuality serving men. Similarly, if we do not challenge the primacy of the body-as-machine metaphor, we effectively ignore the complexities of human experience and misrepresent what it means to 'heal' or to experience dis-ease.

The separation of mind and body in Western medicine is a problem not only from the position of the patient experience, but also because the system is struggling. Today, mental health problems are commonplace, with more and more people diagnosed with depression, anxiety and mood disorders. About twenty-five per cent of people in the United Kingdom are recorded as mentally ill, with women over-represented among that proportion. These statistics may be skewed because men are far less likely to seek help. Additionally, about ten per cent of children are diagnosed with mental health conditions.[64] In Britain the National Health Service, now a pensioner, is in crisis while the numbers, costs, and types of conditions it must treat are increasing.[65] The financial cost of chronic diseases, many of which are classed as 'lifestyle' diseases, exacerbated by the modern Western diet, is especially burdensome. These diseases include chronic obstructive pulmonary disease, Type 2 diabetes, and heart disease.[66] And modern medicine is coming under scrutiny as the drugs that are intended to cure lifestyle diseases (like statins for coronary heart disease) are found to cause additional health problems.[67] There are also large numbers of 'functional conditions', from fibromyalgia to chronic fatigue syndrome, which cannot be explained under the biomedical model. Moreover, these conditions are often gendered and linked to female psychiatric disorders.

The fundamental question that gave rise to this book is whether it is possible to reinvigorate the body with the principles of holism that underpinned humoralism, thus accounting for a person's psychological, social, somatic, and spiritual experience as well as the layered meanings of the body itself.[68] Some of these ideas are developed in the field of integrative medicine, which is considered more below.[69] Seeking a more holistic approach does not mean viewing the body as unchanging or ahistorical. As the chapters in this book demonstrate, our bodies are products of the stories that we tell. Our dilemmas might have been similar to those experienced by Hamlet, but those stories are different. And by taking the body apart, to borrow the architectural metaphors of Renaissance anatomists, we might even be able to construct it anew.

1

Getting it Straight

Spines, Scoliosis, and the Hunchback King

Spine (spīn) n. (*OED*)

1. A series of vertebrae extending from the skull to the small of the back, enclosing the spinal cord and providing support for the thorax and abdomen.
2. The central feature or main source of strength of something; [*mass noun*] resolution or strength of character.
3. The part of a book's jacket or cover that encloses the inner edges of the pages.
4. *(Zoology) & (Botany)* Any hard, pointed defensive projection or structure.
5. (also pay spine) A linear pay scale operated by some large organizations.
6. *(Geology)* A tall mass of viscous lava extruded from a volcano.[1]

During the 2014 Easter holidays I took my children to the Chessington World of Adventures Resort. We queued for an hour for the 'Vampire', a gothic-themed ride through the rooftops, boarding inside a mock gothic abbey complete with dim lighting and dramatic music. Centre-stage was an animatronic organist bent over a pipe organ, his frizzy, cobwebbed head moving in time to the music. The organist's crusty black tailcoat accommodated a large round hump on one side. 'Why are hunchbacks always so creepy?' my teenage daughter asked. 'They are either social misfits or evil. It's not right. They wouldn't be able to do that to people with any other disability.' Millie had more reason to mind than most. At the age of twelve she was diagnosed with severe scoliosis, an abnormal curvature of the spine. Her backbone had not grown straight, but bent twice in an 'S' shape. These twists rotated her ribcage, pressed against her lungs and pushed out one of her shoulder blades.

I had to admit that my daughter had a point. In history, those with twisted spines are either morally dubious, gothically creepy or figures of public ridicule and condemnation; it is a straight spine that suggests, as in the above dictionary definition, the 'main source of strength of something', and 'resolution', even 'strength of character'. Arguably the most famous character in history to have diagnosed spinal 'deformity' was Richard III (see Fig. 1), characterized by William Shakespeare as a grotesque and morally bankrupt hunchback 'rudely stamp'd', 'deformed', and 'unfinished'. Richard III begins the eponymous play with a soliloquy that draws attention to his perceived physical deformities:

> But I, that am not shaped for sportive tricks,
> Nor made to court an amorous looking-glass;
> I, that am rudely stamp'd, and want love's majesty
> To strut before a wanton ambling nymph;
> I, that am curtail'd of this fair proportion,
> Cheated of feature by dissembling nature,
> Deformed, unfinish'd, sent before my time
> Into this breathing world, scarce half made up,
> And that so lamely and unfashionable
> That dogs bark at me as I halt by them.[2]

Shakespeare's Richard III was a man marked by cruelty, as testified by his murder of Edward and Richard, the legitimate heirs to the throne and the so-called 'princes of the tower'.[3] The association of spinal deformity with moral crookedness at the core of Shakespeare's play reflects age-old associations between ugliness and evil on the one hand, and beauty and virtue on the other.[4] While beauty reflected the goodness inside a person, to be as 'ugly as sin' meant the very opposite: ugliness and deformity were equally vilified.[5] Shakespeare's interpretation has influenced most historical criticism of the character of Richard III, though it was published more than a century after the king's death at the Battle of Bosworth. Richard III is not the only famous hunchback; there is also Igor, the crooked assistant of Frankenstein, and Victor Hugo's hunchback of Notre Dame (*Notre Dame de Paris*, 1831), whose Disney incarnation proved the exception to the rule that hunchbacks were evil—though of course the character was too conventionally unattractive to find love.[6]

This chapter explores the history and meanings of the skeleton, especially the spine, and the languages used to describe it. It will start with a more detailed consideration of Richard III, before considering both healthy spines and those affected by scoliosis, a condition that has been relatively neglected in the history of medicine. The spine is usually imagined as the framework of the body, the scaffolding on which everything else sits—a metaphor that, as we will see, has a long history. Perhaps more than any other part of the body the spine is viewed mechanically. Indeed orthopaedics, the medical speciality that deals with bones and their deformities, takes its name from the straightening of crooked bones in youth (derived from the Greek 'orthos' for straight and 'pais, *gen. sg*. paidos' for child).[7]

THE CASE OF RICHARD III

Richard III's remains were discovered within the site of the former Greyfriars Friary Church in Leicester in 2012. Osteo-archaeological evidence suggests that Richard III was not a hunchback, though he did live with scoliosis. His excavated skeleton showed no sign of kyphosis, the condition in which the spine curves outwards and creates the characteristic 'hump' of Shakespeare's hunchback. Leicester University scholars Sarah Knight and Mary Ann Lund, two of the academics charged with dealing with the press after the discovery of Richard III's skeleton in a car park, have convincingly argued that Shakespeare's terminology has clouded judgement of the

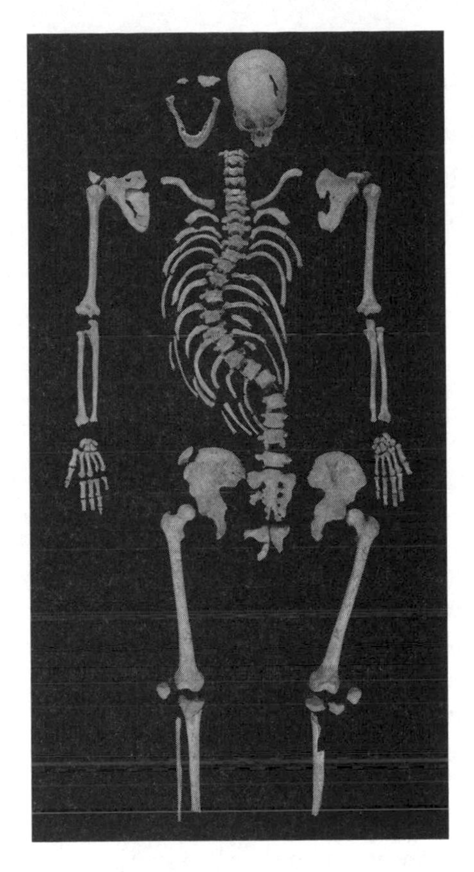

Fig. 1. The skeleton of King Richard III, in which his scoliotic spine is visible.

king's disability.[8] Indeed the discovery of Richard III's skeleton launched a plethora of articles that sought to rehabilitate the king as an impressive and brave warrior rather than a dismal hunchback, who lived with, but was not incapacitated by, scoliosis (see Fig. 1).[9]

Since Shakespeare was writing during the age of Elizabeth I, whose grandfather was crowned after Richard's death, it is perhaps understandable that his character is depicted in a less than flattering light. There is also much art historical evidence that suggests portraits of Richard III were tampered with in the Tudor age, increasing the height of one of his shoulders in one instance, to visually show that the king was physically deformed.[10]

To understand why Richard III was depicted as a hunchback, and the antipathy towards hunchbacks in history, we must consider contemporary attitudes towards disability. Charity towards the weak and infirm has been a staple of Christian tradition since the life of Jesus, but Leviticus made a clear association between physical deformity and moral decrepitude. Certainly no man with a 'defect' was qualified to have a leading role in the church:

> **16** The Lord said to Moses, **17** 'Say to Aaron: "For the generations to come none of your descendants who has a defect may come near to offer the food of his God. **18** No man who has any defect may come near: no man who is blind or lame, disfigured or deformed; **19** no man with a crippled foot or hand, **20** or *who is a hunchback or a dwarf*, or who has any eye defect, or who has festering or running sores or damaged testicles. **21** No descendant of Aaron the priest who has any defect is to come near to present the food offerings to the Lord. He has a defect; he must not come near to offer the food of his God. **22** He may eat the most holy food of his God, as well as the holy food **23** yet because of his defect, he must not go near the curtain or approach the altar, and so desecrate my sanctuary. I am the Lord, who makes them holy." '[11]

The specific wording may have changed slightly from William Tyndale's edition available in the 1530s, but the relevance of the link between physical and moral decrepitude has not.[12] Biblical texts help explain the metaphors that we still associate with the spine, and its association with strength and straightness or crookedness and deception ('spinelessness', of course, being a common term for cowardice). Though figurative, these moral associations dominate the cultural history of the spine, both straight and curved.[13]

SPINE AS FRAMEWORK

The skeleton comprises about 206 bones, several of which fuse together during the ageing process.[14] The spine consists of twenty-four articulating vertebrae and nine fused vertebrae in the sacrum (a triangular bone in the lower back), and the coccyx. The framework of the bones intrigued Renaissance anatomists. In the sixteenth century the Belgian anatomist Andreas Vesalius wrote of the skeleton as part of the core structure of the human being, using architectural language that is still in use today.[15] Professor of Surgery and Anatomy at the University of Padua and later Imperial Physician at the court of Emperor Charles V, Vesalius is commonly regarded as the founder of modern human anatomy, based on the precision and detail of his anatomical plates and his analytical description grounded in dissection and observation.[16]

Vesalius' *De humani corporis fabrica* consisted of seven books or sections, each of which depicted a different system of the body. The first two books were devoted to bones and muscles, with Books 3 to 7 giving an account of soft tissues, including nerves, the vascular system, the digestive and reproductive systems, heart and lungs and brain. The spine was therefore covered in the first book, which dealt with 'Things that sustain and support the entire body, and what braces and attaches them all.' For Vesalius the spine evidenced 'the signal craft of Nature, which fashioned a vertebra in the midst of the back, stable and supported on both ends just as we see builders place one stone between two others in vaulted and arched buildings'.[17]

The skeleton was thus a scaffold on which the entire body's tissues, fibres, and organs rested. Thus the Italian anatomist Niccolò Massa wrote in *Anatomiae Libri Introductorius* (1536) of the bones as 'the foundation of the rest of the parts of the body'. Massa explicitly invoked the work of the ancient physician Galen, who sought to explain the parts of the body in relation to one another, and to construct a set of names for distinct anatomical features in *De ossibus ad tirones* (*On Bones for Beginners*).[18] The bones are 'the hardest and driest parts of the living body,' Galen wrote, as well as the earthiest: 'they sustain and support the other elements of the body as a foundation, for everything is secured and attached to the bones'.[19]

In discussing the 'earthy' nature of the bones, Galen was referring to one of the four elements—earth, air, fire, and water—which along with the four humours composed the fundamentals of traditional medicine.[20] Since humours and fluids governed the body, the condition of its constituent parts depended on its degree of heat. Galen believed that the skeleton was made from the same matter as sperm, as evidenced by its pale colour. The eminent Arab natural philosopher Ibn-Sīnā (known in the West as Avicenna) similarly claimed that bones were derived from the blood, and from dried-up humours. In both the Galenic and the Islamic traditions all the body came from the same matter as sperm, and the bones were 'clothed' with flesh:

> We created humanity from a quintessence of clay, then We made it a drop of sperm in a firm abode. Then We made the sperm-drop into a clot of blood and We made the blood-clot into a lump, and made the lump into bones, and clothed the bones with flesh. Then We made it as a new creation. Blessed be God, the best of creators![21]

GENDERING THE BONES

It was once thought that women had more ribs than men, thirteen sets instead of twelve. This story derived from the biblical tale in Genesis in which God created men and women: 'The Lord God caused a deep sleep to fall upon Adam, and he slept: and he took one of his ribs, and closed up the flesh instead thereof. And the rib, which the Lord God had taken from man, made he a woman, and brought her unto the man.'[22] Against this backdrop of belief, there was some controversy when Vesalius revealed that women and men had the same number of ribs after all. Today it is thought that roughly one in two hundred people have an extra rib, a cervical rib, which connects to the seventh cervical vertebra in the neck. Like a vestigial tail, this is believed to be atavistic: a throwback to the time before hominid evolution.[23]

The German anatomist Samuel Thomas von Soemmering created the first account of a female skeleton, determined like many eighteenth-century scientists to observe sexed difference in every part of the physical body.[24] Similarly, when the French author, chemist, and anatomist Marie-Geneviève-Charlotte Thiroux d'Arconville published images of the female skeleton in 1759 she drew a smaller skull and larger hips than those possessed by a male skeleton.[25] Even within the structures and fibres of the body it was implied that women's hips were designed for motherhood and that the female brain held less capacity for reason than the male. These anatomical proofs of female inferiority were discovered and circulated during the same period in which there was more pressure than ever before for female equality, a trend that continued well into the nineteenth and twentieth centuries. Yet this scientific classification of sexed difference was part of the process by which women were principally defined from the nineteenth century onwards by their reproductive function.[26]

Skeletons continue to be viewed as different in men and women, though all evidence suggests that absolute rules do not apply. Contrary to the determinants around hip and head size proposed in the Enlightenment, some women have small hips and large heads. How peculiar then that we still rely on the shape and size of the skull, and identifying marks on the hip bones, in order to differentiate skeletons by sex.[27] The presumption that women will be smaller has resulted in a bias against physically large women being recognized in archaeological surveys of human remains. Palaeodemographers, those who study ancient human mortality, fertility, and migration through human remains, argue that these Western presumptions (that males are large and females small) make it 'difficult... to recognise older, relatively robust females in skeleton collections'.[28] Similarly, men under thirty who might have thinner, smaller skulls than their older counterparts are sometimes wrongly identified as female.[29]

The diagnosis of scoliosis requires a norm from which it deviates. Few of us have perfectly straight spines. The natural curves of the spine allow it to distribute weight evenly and to absorb impact. Scoliosis is defined by a sideways curvature. According to a seventeenth-century translation of the work of the French surgeon Ambroise Paré, spinal deformities should be understood as follows:

> A dislocated vertebra, standing forth and making a bunch, is termed in Greeke Cyphosis, (Those thus affected we may call, Bunch-backt.) But when it is depressed, it is named Lordosis (Such we may terme, Saddle-backt.) But when the same is luxated to the right or left side, it maketh a Scoliosis (or Crookednesse), which wresting the spine, drawes it into the similitude of this letter S.

Paré's terminology followed that of Hippocrates and Galen. Hippocrates' writing *On Fractures*; *On Articulations*, and *On Places in Man* was dominated by the principles of accurate observation and reasoning. *On the Nature of Bones* is part of the related Hippocratic Collection.[30] For Hippocrates the obvious solution to spinal deformities was mechanical: the physical manipulation of the spine in the opposite direction of the curvature. Hippocrates devised two apparatus to assist him in the treatment of patients: the ladder and the board. In both cases the principle was to pull the spine manually back into a straightened position with minimal damage to the spinal cord. To achieve a curve reduction, the patient's ankles would be tied to a wooden ladder and further bands were added above and below the knees and also at the hips. The hands were tied to the body, and then the ladder was lifted 'against some high tower or house-gable' against which the patient would be manually shaken—a painful-sounding process that Hippocrates termed 'succussion':

> The ground where you do the succussion should be solid, and the assistants who lift well trained, that they may let it down smoothly, neatly, vertically, and at once, so that neither the ladder shall come to the ground unevenly, nor they themselves be pulled forwards. When it is let down from a tower, or from a mast fixed in the ground and provided with a truck, it is a still better arrangement to have lowering tackle from a pulley or wheel and axle.[31]

Today Hippocrates' technique represents a type of 'lumbar traction' that still has its place in physical therapy.[32] Hippocrates' second technique using a board worked in the same way, with straps, wheels and axles to enable the traction. It was adapted by Galen, who added the use of pressure in his work *On the Usefulness of the Parts of the Body*, which provided an analysis of spinal anatomy that varies little from modern-day texts. Galen detailed four separate conditions to the spine that was not straight: kyphosis (when the spinal column moved backwards), lordosis (when it moved forwards), scoliosis (when it moved to the side) and succussion (where there was no spinal deformity but the invertebral articulations had still moved).[33]

Both Hippocrates and Galen attributed spinal deformities like scoliosis to 'gatherings' or abscesses on the spine, as well as to the postures that might be adopted by patients in bed. This is compatible with a humoral view of the body being influenced by all the non-naturals, including movement and/or lack of movement and posture. It is also compatible with modern physiotherapeutic techniques that attempt to reduce spinal curvature through close attention to posture and the engagement of symmetrical muscle control.[34] Emphasis was placed on diet as well as exercise by Galen and, despite the mechanical nature of the treatment, on the whole person as well as the spine. Today nearly eighty per cent of cases are deemed idiopathic, meaning that they have no known cause, though scoliosis can be often linked to other conditions such as cerebral palsy. The condition is more common among girls than boys, again for unknown reasons, though this was not noted in the writings of Galen or Hippocrates.[35]

Traction techniques, for all their problems, are largely non-invasive. They are designed to halt pathological growth and development, rather than realigning the spine itself—attempts at which were often considered too risky. Bracing technologies were implemented from about 650 CE, when the Byzantine Greek physician Paul of Aegina (625–690) bandaged scoliosis patients with wooden strips.[36] In 1575 Paré created the first metal scoliosis brace intended to be worn full time to correct the growing spine.[37]

European physicians and writers showed increased interest in spinal deformities from the seventeenth century, as well as perplexity over its cause. Suggestions included faulty posture and misuse of stays and corsetry, muscular imbalance, post-fall trauma, constitutional defects, menstruation, running, and even standing on one leg.[38] There were written accounts of scoliosis by the Dutch anatomist Antonis Nuck, the Swiss physician Jean-André Venel, and the German physician Johann Georg Heine, all of which relied on mechanical corrections. Nuck used a head suspension appliance called the torques. Venel treated patients with an anti-gravity bed at night and a corset during the day, and Heine founded the first orthopaedic institute in Germany in 1816. There were significant similarities between the techniques, and in each case the greatest clinical risk was spinal cord damage and paralysis.[39] In 1764 the first mechanical bed for children with scoliosis was created by François Guillaume Levacher de la Feutrie, who presented it to the Académie Royale de Chirurgie de Paris. The idea behind the bed was that children's bones were malleable and early correction more likely to have better results. However, like other forms of traction the bed proved ineffective and had mainly fallen into disuse by the early 1800s.[40]

Treatment in the UK has historically been more conservative than in France. Though doctors were interested in spinal deformity it was more than two centuries before its treatment was established: until then it was the province of surgeons, mechanics, masseurs, and bonesetters.[41] In 1837 London's first infirmary for spinal diseases was established by the physician Edward Harrison, ten years after the first publication of his *Pathological and Practical Observations on Spinal Diseases*.[42] Harrison never referred explicitly to scoliosis, preferring to discuss spinal 'deformities' as a whole. His therapeutic techniques were conservative; initially he depended on 'recumbency, frictions, daily pressure and slips of adhesive plaster'.[43] This was followed by elaborate bandaging of his patients, including a 'stuffed wooden shield' placed on the back, turtle-like, in order to further increase the pressure on the spine.[44] Finally he developed a steel machine, 'constructed on the principle of a windlass', a machine for moving heavy weights, 'to draw out the spinal chain, and place the vertebrae further apart from each other'.[45] These bracing techniques adapted the corset that had been developed by Paré three hundred years earlier.

Paré's bracing was based on the belief that all spinal deformity resulted from its dislocation. Paré attributed scoliosis to trauma, poor posture, and in girls the use of bodices and the requirement to curtsey. For adults with scoliosis he recommended the Hippocratic treatment that included forcible horizontal traction, but for children he prescribed brace or iron corsets:

> With holes in so that they are not too heavy, and they will be so well fitted and padded that they will not cause any injury. They would be changed often if the invalid does not achieve the three dimensions. And for those girls who grow, it would have to be changed every three months, more or less as necessary: for otherwise, instead of doing good, it would do harm.[46]

Despite some modifications in the material used, developments in the treatment of scoliosis stagnated until the end of the nineteenth century. Lewis Albert Sayre, a leading American orthopaedic surgeon and subsequently president of the American Medical Association, recommended gymnastic-style exercises to strengthen muscles on the weaker side of scoliotic spines. He also developed a traction technique that he used along with plaster casting. The patient was suspended off the ground by supports at her chin and underarms before the plaster jacket was attached. This was to be removed at night and during exercise but kept in place at all other times.[47]

The Milwaukee brace was invented in the twentieth century: a full torso brace that extended from the pelvis to the base of the skull.[48] Three bars, two posterior and one anterior, were attached to a pelvic girdle made of leather or plastic, as well as a neck ring. This brace was normally used with growing adolescents and worn for 23 hours a day for several years, or even permanently. A related brace is the Boston Brace, an underarm brace made of solid plastic with buckles that tighten to press internal pads against the spine. The jury is still out on the efficacy of bracing.[49] A medical review of the records of over 1000 patients fitted with the Milwaukee brace between 1954 and 1979 found that the higher the curve in the first place the more likely a child was to undergo surgical correction.[50] However, there is a lack of systematic analysis of scoliosis treatment and such studies have been criticized, partly because patients are rarely followed throughout their lives. We do not know the impact of ageing on the scoliotic spine, nor how the curve might progress long term once the brace is no longer used. There is no data to prove that bracing or exercise or any extrinsic factor can permanently alter the course of scoliosis.[51] Moreover, there are social as well as medical reasons why bracing might be ineffective, especially amongst adolescents. Lack of compliance is a problem, either because of the appearance of the brace for image-conscious teenagers, or the discomfort (rubbing, chafing, and digging in) of the brace into the skin of the wearer. Set against the limits of the brace's proven efficacy, these psychological impacts matter.[52] After all, the best a brace can do is to halt a curve. It cannot rectify it. There is even a possibility that its use may weaken back muscles, thus creating a negative effect overall.[53]

SURGICAL INTERVENTION

From the late nineteenth century technological developments have allowed the course of scoliosis to be viewed and measured from a distance. Seeing inside the body without cutting it open has been identified as one of the most transformative moments in the history of modern pathological medicine, though it is likely such technologies were initially far more dangerous to patients than scoliosis itself.[54] Wilhelm Conrad Röntgen's 1895 discovery of X-ray radiation enabled physicians to study skeletal anatomy without dissection.[55] During the early 1900s, however, spinal radiographs required long exposure times that often resulted in poor quality spinal radiographs because of patient movement. Though X-ray technologies spread quickly, there was no straightforward diagnostic interpretation of those images.[56] Moreover, the negative health effects of X-ray radiography had not yet been discovered. There are still questions about the number of X-rays taken during the course of an individual's treatment. In the case of adolescent onset of scoliosis this is particularly problematic because of the sheer quantity of X-rays needed in the 'watch and wait' medical approach.[57]

Measuring the extent of spinal curvature today is done using the 'Cobb Angle' assessment, so named after the American orthopaedic surgeon John Robert Cobb.[58] Surgeons make two lines on an X-ray based on the position of the vertebrae. They locate the vertebra at the top of the curve with the most tilt, and then draw a line parallel to the top of the upper or superior end plate. Then they find the vertebra at the bottom of the curve that is tilted the most and draw a line parallel to the bottom of the inferior end plate. The angle where these lines intersect is the Cobb angle.[59] In the twenty-first century, surgery is recommended for curves with a curvature above 45 degrees according to the Cobb angle; curves that would be cosmetically unacceptable in adulthood, that cause pain, or that when combined with other health problems like spina bifida or cerebral palsy, interfere with sitting and possibly such basic physiological functions as breathing.

Scoliosis surgery, also known as spinal fusion with instrumentation, is major surgery. There are two procedures involved, each of which is distinct. One aspect of the surgery involves fusing the vertebrae, using transplanted bone from another part of the patient's body or from a donor, though synthetic bone can also be used.[60] There are two main ways that surgery is carried out: anterior or posterior fusion. In the former the incision is made at the side of the chest wall and in the latter it is made at the rear. The second part of the procedure is the insertion of metal hardware to support the correction. Since the 1950s a metal system of straightening has been in place. In 1955 the American orthopaedic surgeon Paul Randall Harrington developed the Harrington rod, the first surgical device for straightening and immobilizing the spine within the body.[61] The Harrington rod was used to correct more than one million cases of scoliosis. It consisted of a stainless steel rod attached to the spine at the top and bottom of the curve with hooks. Ratchets were then tightened to straighten the spine. Finally bone from the patient's own hip, and later donor bone, was used to stimulate a fusion between the vertebral spaces.

The main shortcoming of the Harrington method was it could not produce proper alignment between the skull and the pelvis. Nor could it address rotational deformity; scoliosis affects the spine not only vertically but also horizontally, because the spine and ribcage rotate as the spine bends. This was the cause of my daughter Millie's back 'hump' (as well as the historical suggestion that Richard III was a hunchback), as the rotation of the ribcage pushed her shoulder blade up and out like the beginnings of an angel's wing.[62] Because of the limitations in the Harrington method, it was possible for unfused parts of the spine to compensate for the fused parts, leading to excessive wear and tear, arthritis, muscular stiffness and ultimately disability.

Modern methods involve a combination of more flexible techniques to fix the spine, though the process remains largely mechanical. An incision is usually made in the back during posterior entry; the length of which depends on how many vertebrae the surgeon intends to fuse. The surgeon opens the incision and strips away all tissue and muscle from the spine. She then removes the facet joints from between the vertebrae, enabling the manipulation of the spine prior to fusion. In Millie's case, her ribcage was rotated back to the centre before her spine was reinforced with two sterile cobalt metal rods that were welded into place by fourteen titanium screws. Donor bone and coral grafts (one of the closest natural elements to human bone) were added to Millie's spine, creating a reaction in which the vertebrae began to fuse together.[63] This fusion process normally takes up to a year, though it will be another two years before we might contemplate another trip to Chessington.

The degree of flexibility after spinal fusion depends on the nature and degree of the original curve and the number of joints that are fused together. Millie's spine was fused from T2 to T12; that is from the second to the twelfth vertebrae of the thoracic spine. The metal of the rods responds to the weather; on cold days they make her back feel stiff. She can feel the top of the rods through her skin and her back is a mass of painful sensitivity: the nerves and muscles and blood vessels that were sliced through for the surgeon to access her spine are slowly growing back, accepting the cobalt chrome and titanium invaders as though they are part of her, which of course they are. They help her to sit straighter, to walk taller, and to wake up most mornings without pain.

Some of the questions that my daughter has asked during her recovery has made me think more deeply about the meanings of any kind of prosthesis surgery or implants.[64] At what point do the rods and screws become part of her 'real body'? Does she still have scoliosis though the twist in her spine has been forced into submission? The first few times she stood after surgery her body's memory tilted her sideways because that was her spine's version of 'straight'. Stumbling and falling were all too real dangers until her body adjusted to the tucked-in shoulder blade, the de-rotated ribs and the extra two inches of height. Of course Millie is less routinely aware of her new spine in the same way that she might be if she had acquired a replacement heart or she could feel the air move in and out of a new set of lungs. But she is more aware of her spine than most, for sometimes it clicks and jolts and reminds her that it is still knitting itself into the fibre of her being.

How Millie self-identifies as a post-operative person is, for me, an important and not always easy insight into the challenges of medical intervention and pathologization. Of course Millie is not reducible to her parts, though this certainly was the *de facto* presumption through her journey as a patient: through the processes of consultation, evaluation, diagnosis, prognosis, surgery, recovery, and beyond. She was entered into the hospital system and given a number before being photographed, measured, weighed, and scanned with X-rays and Magnetic Resonance Imaging (MRI) technologies over a period of three years, before undergoing spinal fusion. Routine surgical surveillance will continue for the next few years, with her spine being the sole focal point of discussions as we sit around the surgeon's desk, straining to see the black and grey images on his computer screen. Millie's spine with all its attachments is most definitely part of her whole: not her body as opposed to her mind, not the framework on which her flesh and muscles have been arranged, but her embodied self, the collection of thoughts, dreams, fears, and fibres that make her who and what she is; just a stronger, straighter version.

We are grateful to my daughter's surgical and nursing teams at Great Ormond Street Hospital. Her everyday discomfort has been profoundly reduced, and she is getting used to being straight. But the division of the body into parts, and the separation of the mind from the body as a disciplinary, training, and economic imperative means that the emotional effects of Millie's surgery were not considered. In an ideal world, moreover, we would know *why* the scoliotic spine chooses to curve and bend, as well as how we might prevent it. Modern surgical techniques are fundamentally about correcting a mechanical structure, in much the same way as Hippocrates practised with his ladder. And the hard plastic brace that Millie wore night and day for six months, making her look and feel like a human mannequin, was based on the same principles as the one invented by Paré.

In modern health care our ability to treat the spine as a part we can repair, like the heart, like a limb, necessarily displaces our 'self' from the body part in question. It disengages the physical from the psychological, placing less emphasis on a patient's embodied experience than might otherwise have been the case; certainly less than would have been the case in the ancient humoral tradition. We were fortunate that Great Ormond Street

Hospital was able provide psychological counselling to young people undergoing surgery, though we had to ask for it as a separate resource. Psychological support will not be an option at many hospitals in the UK as a result of over-stretched resources and budget prioritization.

The lack of connection between spinal and psychiatric medicine is particularly striking when one considers that the spine was once linked to a range of psychological and psychiatric conditions, including concepts of 'neurosis' and early forms of Post-Traumatic Stress Disorder (PTSD). The nineteenth century was replete with ill-defined nervous complaints like 'neurasthenia'—which was characterized by fatigue, headache, and irritability associated with emotional disturbance—and 'nervous breakdown', which is not a medical term but nevertheless remains a popular one to describe feelings of emotional and physical overwhelm. With the expansion of the railways and the arrival of railway accidents in the Victorian age, a new form of spinal disease emerged—'railway spine'—in which the nervous system was similarly injured and weakened. After being involved in railway accidents, people reported insomnia, feverishness, palpitations, and a whole range of weaknesses associated with nervous complaints. Thomas Buzzard, physician to the National Hospital for the Paralysed and Epileptic in London, suggested that the jarring and jostling of the railway carriages were causing damage to passengers' spines that subsequently impacted on their psychical health.[65] The rationale for this belief was grounded in eighteenth-century ideas of spinal irritation discussed by the Scottish physician Robert Whytt, for whom the spine had considerable autonomy from and influence on the brain.[66]

Today we do not think of the spine as an autonomous structure, though we do allow agency to organs other than the brain. I am not advocating the return of 'neurasthenia' as a recognized condition than connects psychological and physical symptoms. What I am identifying is a process by which the separation of the mind and body into parts and systems necessarily means that spinal conditions like scoliosis are viewed as structural and functional problems addressed exclusively within orthopaedics. There have been no significant evidence-based investigation of other non-surgical options, including bracing and physiotherapy and even acupuncture, though a 2008 pilot study on the latter did indicate potential benefits.[67] There is evidence that public attitudes are turning towards a more holistic form of medical treatment, but the funding for research (and the necessary proof of efficacy) is not widely available.[68] The same is true of potential links between thyroid and hormonal, and other deficiencies in children and the subsequent onset of scoliosis.[69]

These claims are unproven and look set to remain that way for the foreseeable future. After all, specialisms like orthopaedics and endocrinology do not tend to work together for reasons that are practical, institutional, and financial. Both disciplines are concerned with quite different structures and systems in a body that is entirely divisible and material. We have developed surgical responses to spinal conditions like scoliosis and in this alone we have moved on from ancient therapeutics. All other forms of treatment, and attitudes, are remarkably unchanged—including perhaps our response to disability. In March 2015 Richard III was accorded a royal burial, though not everyone was happy about it.[70] For many people he is best remembered not only as the murderer of two young princes—a claim that has not been proved—but also as Shakespeare's 'Hunchback King'.[71] Our attitudes towards spinal 'deformity' as a physical, emotional, and social phenomenon clearly have some way to go.

2

Beauty and the Breast

From Paraffin to PIP

'When I had the implants put in, I would get wolf whistles when I walked down the street... I truly believe women should be free to choose. But, to be honest, there are times when I think I would like to have mine taken out... I started to get pain in the Eighties and somctimes it lasts for five to six weeks. It feels like I've broken a rib.'[1]

The French man of letters Anatole France—novelist, poet, and winner of the Nobel Prize for Literature in 1921—is said to have observed that 'a woman without breasts is like a bed without pillows'. This is a challenging reduction of the female body to an object that is comfortable and useful while being (a) as domestic and controllable as a home's soft furnishings, and (b) incomplete if a woman is flat-chested, whether because her breasts do not develop during puberty, or from breastfeeding, or even from disease. France's words should be ridiculed, and yet they seem to strike at the heart of a wider, broadly acknowledged truth about attitudes towards women's bodies in general and breasts in particular. Breasts symbolize womanhood: they are sexualized, nursed at, revered, mutilated, sniggered at, and politicized, for how they look and feel, as well as for what they do.[2] Breasts are both sexual organs and a source of nourishment, which shatters the careful barrier we have erected in the West between motherhood and sexuality.[3]

This cultural obsession with breasts, with their size and shape, has direct implications for women's health. The most obvious example is in cases of cancer; perhaps the most feared disease in the Western world and a space where ideas about personal responsibility, disease, technological innovation, and ethics coalesce around the female breast.[4] Consider the 2013 media storm surrounding the actor Angelina Jolie when she elected to have a double mastectomy after genetic testing revealed a high cancer risk.[5] Not only did the 'Angelina Jolie effect' inspire many thousands of ordinary women to ask their GPs (General Practioners or NHS nursing teams) for genetic testing, and open up the broader question of breast surgery as a prophylactic measure, it also raised the question of silicone breast implants: Should you? Could you? Would you look 'normal'? Is it 'safe'?[6] Unwittingly or otherwise, Angelina Jolie became an icon for women's right to take control over their own health and their own breasts.[7]

Jolie is said to have had silicone implants in place of her excised breast tissue. Not all silicone implants are inserted because of cancer risks or reconstruction. A recent study by the non-profit organization the Institute of Medicine (IOM) found that of the 1.5 million American women with silicone breast implants, more than two-thirds chose implants because they were unhappy with the size and shape of their breasts.[8] In other words, they underwent 'augmentation' as an elective procedure. In British and North American culture, big breasts are a symbol of sexual allure and individual attractiveness and more democratically available than ever before due to cosmetic procedures.[9] So much so that *The Huffington Post* announced 2012 the 'year of the silicone breast implant', breast implants being the most popular cosmetic procedure in the United States (US) according to the American Society for Aesthetic Plastic Surgery (ASAPS).[10] A similar pattern is found in the UK where over 50,000 cosmetic procedures were performed in 2013. According to the British Association of Aesthetic Plastic Surgeons (BAAPS), cosmetic surgery was steadily on the increase despite the recession, with a rise of seventeen per cent in surgical procedures since 2012. The value of UK cosmetic procedures as a whole was £2.3 billion in 2010, estimated to increase to £3.6 billion by 2015.[11]

This trend raises important philosophical and practical questions about the motivation of those who seek implants. This chapter explores those questions, and asks why so many women are so unhappy with their breasts that surgery seems the only recourse. The terms 'plastic' and 'cosmetic' surgery are often used interchangeably, but the two branches of surgery evolved separately. It is important to distinguish between reconstructive and cosmetic (plastic) surgery from that which is purely aesthetic (cosmetic). I have chosen to focus purely on cosmetic surgery because the pursuit of an objective beauty ideal affects women in particular. The drama of breast augmentation also opens up feminist and ethical issues around consumer choice and rights, medical safety, and ethics that are not normally debated in the histories of medicine and the body, and are also peripheral concerns in the histories of breast surgery, which focus on surgical prowess and more generally on the cult of beauty.[12]

With the notable exception of the 'nursing Madonna', or Madonna Lactans, the ideal breast was an unused one: 'small firm and spherical', especially in the Classical Greek and Renaissance traditions.[13] In 1840 the English surgeon and anatomist Sir Astley Paston Cooper published a treatise in which he considered the aesthetic, healthy, and pathological functions of the female breast.[14] Before describing its morbid anatomy, Cooper explored the purpose and appearance of female breasts. They are first and foremost for 'suckling', he wrote, placed where they were so that the infant might receive both nourishment and the mother's 'tender and regular affection'.[15] Traditionally, the task of nursing an infant was farmed out to others, at least among the wealthier levels of society, so Cooper's comments reflect both a change in cultural practice (breast feeding from the natural mother being the preferred choice from the nineteenth century, though according to distinct socio-economic and geographic differences), and the moralizing discourses attached to breastfeeding.[16] Our modern, cultural ambivalence towards breastfeeding as both ethical necessity and social embarrassment is not new.[17] Nor is our dual interest in the breast as both a sexual and aesthetic object, though there was a heavily racialized prejudice about the perfect breast. As Cooper rhapsodized:

> The breasts, from their prominence, their roundness, the white colour of their skin, and the red colour of the nipples, by which they are surmounted, add great beauty to the female form for prior to the age of puberty, the girl and boy differ but little in the shape of the chest, or in its general appearance; but as the breasts develop, the female figure is established in all its elegance.[18]

In Africa, by contrast, especially among the Hottentot women:

> Their breasts hang by a fold of skin, very loosely upon the abdomen, as a stone does in a sling... this great relaxation of the breasts is not peculiar to the females of warm climates, but is also seen in the coldest regions which man can inhabit. The Esquimaux [Eskimo] women, who live in cabins excessively heated through a long winter, are, I am informed, subject to similar changes as those of hot climates, their breasts becoming very pendulous.[19]

Shakespeare in Sonnet 130 satirizes fashionable physical perfection and rejects the ideals of lead-painted beauty that made 'snow'-coloured breasts more attractive than 'dun'.[20] By the Victorian period those ideals were popularized by Darwin's doctrine of sexual selection.[21] Of course there had always been ideals of the beauty of the female form and the female nude has long been an object of artistic culture.[22] What the history of cosmetic surgery reveals is the emergence of a different kind of dynamic: away from the ideals of classical art, the female body has emerged not merely as an abstract but a realizable ideal; it has fallen to surgeons to provide what nature did not.

Between the 1950s and the present day the concept of psychological 'need' emerged in relation to breasts that were perceived as too small. That need was met by the evolution of plastic surgery techniques and materials, the development of specific standards of female beauty and the growth of a mass market in surgery as a consumer choice. Silicone implants also underwent several transformations of their own: from pathological prostheses to coveted consumer objects to walking or ticking 'time bombs'.[23] An exploration of the evolution of silicone implants reveals a startling disjuncture between the narratives of suffering on the part of implant recipients and of the evaluation of risk by a medico-scientific community invested in the continued use of silicone. The existence of these disputes raises important ethical questions about the ways women's bodies have become a site of conflict for competing interests—not only of manufacturers, surgeons, and academics, but also of the media which simultaneously criticizes women's bodies while questioning their right to choose—or at least to have those implants paid for by the NHS.[24]

THE ORIGINS OF COSMETIC SURGERY

Derived from the Greek plastikos, which translates as 'to mould', plastic surgery has been identified in ancient Indian Sanskrit texts that describe procedures to repair noses and ears lost as punishment for crimes such as adultery. In 600 BCE the Indian surgeon Susruta, whose work was compiled as *Susruta Samhita*, described a method of rhinoplasty that used skin from the cheek or the forehead of the patient.[25] By the first century CE, the Romans were also undertaking a range of aesthetic surgical procedures, such as breast reduction and circumcision.[26]

Since the time of Galen, surgeons have attempted to correct eyes that drooped and noses that were considered misshapen. Yet the 'father' of modern plastic surgery is usually said to be Gasparo Tagliacozzi, who worked in Bologna, Italy, in about 1590. His technique was to transfer a skin flap from the arm to the nose, an intervention necessitated by the number of street brawls and duels that took place during the period.[27] It is likely that some of his patients were also men who had succumbed to some type of nose excision as a punishment; it was particularly emasculating for a man to have his nose sliced in two or removed entirely.[28]

The predominantly reconstructive nature of plastic surgery became established as a result of the World Wars. Surgeons came into their own as they dealt with the aftermath of trench conflict.[29] New techniques were required to deal with the carnage of the battlefields, and surgeons had to work together quickly. The weapons of war, high-explosive shells, machine guns and poison gas also created huge numbers of casualties, many of whom needed rehabilitative surgery to repair limbs and enable skin grafts. The trench warfare of the First World War encouraged the development of plastic surgery, as soldiers needed treatment for shattered jaws, skull wounds and demolished faces. Harold Gillies (later Sir Harold), a New Zealand otolaryngologist working in London, developed many of the techniques of modern facial surgery in caring for soldiers with disfiguring facial injuries.[30] Gillies acted as a medical minder in the Royal Army Medical Corps, supervising a French-American dentist who was not allowed to operate unsupervised, but who was attempting to develop jaw repair work. After working with the renowned French surgeon Hippolyte Morestin on skin grafts, Gillies and the army's chief surgeon Sir William Arbuthnot-Lane established a facial injury unit at the Cambridge Military Hospital in Aldershot and later a new hospital in Sidcup.[31]

When the Second World War broke out, plastic surgery provision was divided between the different services of the armed forces. Gillies worked at Rooksdown House near Basingstoke, which became the principal army plastic surgery unit. Gillies' cousin Archibald McIndoe (also subsequently knighted for his services) worked with Gillies before founding a Centre for Plastic and Jaw Surgery in East Grinstead.[32] McIndoe developed new techniques for treating burned faces and hands and skin grafts, as well as innovative methods to integrate ex-soldiers back into the community. By the end of the Second World War, plastic surgery covered the reconstruction of all areas of the body, and led to the development of a number of professional organizations. The British Association of Plastic Surgeons was founded in 1946, now named The British Association of Plastic Reconstructive and Aesthetic Surgeons (BAPRAS).

The activity of plastic surgeons and the development of cosmetic surgery as a specialty gave it a rationale and an impetus that continued long after the end of the Second World War. The British Society of Aesthetic Plastic Surgeons (later the British Association of Aesthetic Plastic Surgeons or BAAPS) held its first meeting in 1979. In North America, professional associations were formed earlier. The American Society of Plastic Surgeons (ASPS) was formed in 1931, and the American Board of Plastic Surgery was established in 1937. The American Society for Aesthetic Plastic Surgery (ASAPS) was not founded until 1967.

Plastic surgery on the breast was initially reconstructive, in keeping with how such surgical methods had evolved. In the main, that meant that prostheses were used to enhance a woman's breast after invasive surgery or tissue removal. Implants went through a wide variety of permeations with varying degrees of success: from lipoma autotransplantation (the movement of fatty cells from one part of the body to another), paraffin injections, ivory and glass balls, ground rubber, ox cartilage, polyethylene chips, polyurethane sponge, silastic rubber, and liquid silicone.[33] In the late nineteenth century the German-Bohemian surgeon Vincenz Czerny, head of the surgical departments of the universities of Freiburg and Heidelberg between 1871 and 1906, published his first account of a breast implant carried out to avoid asymmetry after he removed a tumour from a patient's breast. The implant was made from tissue harvested from the patient's abdomen.[34]

As a reconstructive process the moral and ethical bases of plastic surgery were straightforward. Surgical reparation was part of an individual's rehabilitation into the social world, and it explicitly engaged with questions about appearance, body image, and mental health as an aspect of healing.[35] The situation was more problematic in the case of purely cosmetic surgery, in which an individual perceived there to be something 'wrong' or lacking about his or usually her physical appearance, and sought intervention through an elective surgical procedure—usually performed privately and as a response to internalized ideals of bodily perfection.[36]

HYPOMASTIA AND THE PATHOLOGIZATION OF SMALL BREASTS

Despite early experimentation, medical interest in electively increasing the size of a woman's breast was rare until the mid 1930s. This is unsurprising when one considers the fashion in women's bodies at the time. Historians of fads and fashions in female appearance have shown how the 1920s aspirational look was 'skinny—no hips, no breasts, just a straight figure'.[37] Cosmetic surgery was reserved for overly large or 'hypertrophic' breasts, which might also impact on the respiratory, circulatory, and locomotor systems.[38] However, some surgeons were experimenting with fat transplants to alter the shape of women's bodies in the late 1920s, which included moving fat from the buttocks or abdomen to the chest. This process of 'fat transfer' was used experimentally for a wide range of purposes.[39] There was already a professional distinction developing between those who worked in 'reconstructive' surgery, and those who focused primarily on 'cosmetic' concerns—as well as a degree of rivalry.[40]As the American surgeon John Staige Davis put it, 'true plastic surgery' ('plastic' in the traditional sense of moulding and shaping) was absolutely distinct and separate from what is known as 'cosmetic or decorative surgery'.[41]

The ideal of a larger bust size peaked during the post-war years, arguably as a reaction to the slenderness of 1920s fashion and to the paucity imposed by rationing. The aspirational bodies of Hollywood in the 1940s and 1950s were Marilyn Monroe, Jane Russell, Lana Turner, and Sophia Loren, all of whom were famous for their hourglass figures. The gamine look, as it would be popularized by Audrey Hepburn and Twiggy, was not yet as fashionable as breasts that pushed out college sweaters and defied gravity. The Hollywood movie industry set increasingly unattainable norms and conventions around the female form.[42]

In the 1940s women who failed to conform to the ideal were said to have bemoaned their fate to their surgeons: 'Doctor, why can I not have a breast which makes me look as good as any other woman, if not better?' was the alleged cry of female patients according to cosmetic surgeons.[43] The 'need' was there, they insisted, though surgical treatment was not yet available. In the 1930s the cosmetic surgeon H.O. Bames described three kinds of 'anomalous' breasts: abnormally large and pendulous; normal in size but prolapsed, and abnormally small (the latter of which he viewed as an endocrinal rather than a surgical problem, requiring hormonal intervention).[44] By 1942 the possession of 'infantile' breasts was classified as 'hypomastia'–or underdevelopment–by the Hungarian surgeon Max Thorek in his 1942 textbook.[45] Thorek did not suggest any surgical corrections at this point but variations in breast size and shape were increasingly medicalized. Post-war plastic surgery journals showed a consistent interest in breast augmentation. By 1950 Bames was writing not of endocrinal solutions, but of surgical ones to 'breast malformations and a new approach to the problem of the small breast'.[46] No endocrine therapy had worked, he explained, 'hence the only recourse we have for their solution is Plastic Surgery'.[47]

Thus cosmetic surgery came to the rescue of women with an identified need for larger breasts, in a narrative of surgical advance. The surgical profession had grown in numbers, training, and technical proficiency. In the increasingly youth-dominated culture of the West, they found a strong market among middle-aged, middle-class women with disposable income and all the 'pathological' body parts that accompanied child-bearing, including breasts that were too small, shrunken, or insufficiently perky. As the number of hair salons, manicurists, and beauticians escalated, 'beauty contests' were established and mass consumerism became part of the 'cult of the body beautiful', dominating post-war Britain and North America.[48] Modern day sociological concerns—about the impact of the media on the self-esteem of women and girls, the existence of aspirational ideals and tools for self-improvement and transformation—first originated in the post-war period.[49]

Indeed it was the writings of the Austrian psychotherapists Alfred Adler and Sigmund Freud, with their emphasis on the 'inferiority complex' and concepts of the self in ego development, which became synonymous with the pursuit of physical perfection.[50] Of course that was not the aim of Adler and Freud, but looking good—or looking 'right'—was transformed into a psychological necessity; self-esteem and self-confidence irredeemably crushed if one possessed sticking-out ears or a flat chest. Over the same period women's magazines and popular psychology stressed both the attainability and the obligation to pursue one's 'best self', and cosmetic surgery rapidly arose as a route to self-improvement.[51]

In this context 'hypomastia' became a psychiatric as well as a physical condition. It was no longer just a label to describe breast size but also described the accompanying feelings of inadequacy, shyness, neuroses, frigidity, and depression. 'Literally thousands of women, in this country alone', cosmetic surgeons and psychiatrists from Johns Hopkins reported in 1957, 'are seriously disturbed by feelings of inadequacy in regards to concepts of the body image... "augmentation" mammoplasty (for the small breast) is usually requested by patients with emotional problems'.[52] The positive impact of surgical alteration could be seen to outweigh any potential physical risks of surgery.[53] Psychological need was pitted against those who viewed breast augmentation as a physically hazardous operation.[54]

Two of the most important considerations for cosmetic surgeons were aesthetic appearance and touch: what combination of technique and material would provide the most malleable, secure, and 'life-like' alternative to human breast tissue? Without regulation the professional associations worked this out through trial and error. Surgeons collaborated with the women who were said to drive the market and the manufacturers who supplied the implants. During the 1950s a range of new techniques and tools became available, including synthetic materials like polyvinyl alcohol and polyethylene. Cosmetic surgeons injected liquid substances straight into the breast tissue, just as they had done with fat, or moulded implants that could be inserted beneath the chest muscle. Robert Alan Franklyn, a Hollywood surgeon, implanted 'surgifoam' in women's breasts (a kind of absorbable gelatin substance sheathed in teflon) as well as experimenting in the subcutaneous injection of liquid silicone through a technique called 'Cleopatra's needle'. Other surgeons experimented with *Ivanol*, a polyvinyl substance related to plastic.[55] Synthetic implants gave a better overall result than injections, largely because the body could reabsorb injected substances, or those substances might create ripples and lumps beneath the skin.

THE SILICONE AGE

It is against this backdrop of defined need, greater surgical skills, and more experimentation and synthetic materials that we need to situate the medical experience of Timmie Jean Lindsey, cited in the epigraph, and the emergence of the silicone implant in North America. It is telling, given the subsequent furore over the health and safety of silicone implants, that there is little about Lindsey in the academic or scientific literature. There is a perennial problem of women, and of patients, being excluded from the historical record. Lindsey's name appears as a footnote in a handful of papers, but she is otherwise anonymous, her experience not part of the official record.[56] Only with media interest in her case fifty years after her implants, as part of a broader historical retrospective, has her voice been heard.[57]

The official narrative of the first breast implant is well known in the medical literature on breast augmentation.[58] In tone it smacks of the locker-room and the boys' club; unfortunately not unusual rhetoric for the language of scientific discovery.[59] In 1962 the American plastic surgeons Thomas Cronin and Frank Gerow were working with scientists from the Dow Corning Center for Aid to Medical Research on improving and refining breast implants. Dow Corning was an amalgamation of a glass works and a chemical company. By that time a number of different materials had been used as implants, mostly plastic, sponge-like materials like Polystan (polyethylene tape and polyethylene) that produced a result 'pleasing to the eye', but hard and 'rocklike' to the touch.[60] The inspiration for the first silicone gel implant is said to have been a blood transfusion bag. One of Cronin's employees, Thomas Biggs, recounted to an in-house publication the events that would later become famous. During the silicone gel implant's developmental stages, Cronin apparently visited a blood bank, and 'upon feeling the new and improved flexible plastic bag that contained the blood, he observed that it felt like a breast ... The rest, as they say, is history'.[61]

Inspired by this haptic revelation, Cronin and Gerow used silicone to make a thin, flexible bag and also a gel-like substance with which to fill it. Silicone is derived from silicon, a semi-metallic or metal-like element that in nature combines with oxygen to form silicon dioxide or silica.[62] Additional processing can convert the silicon into a long chemical chain or polymer, which can be liquid, gel, or solid. Today silicone is widely used in the medical and cosmetic industry, used in lubricants and oils, as well as suntan lotions, soaps, antiperspirants, and chewing gum. As early as the 1950s, scientists debated the safety of silicone and warned against its widespread use without proper testing.[63] Cronin and Gerow implanted one of the prototypes in the body of a dog. They were jubilant that the new 'natural feel' silicone implant had arrived: 'the dog was fine, so they implanted it in a person and she got along just great'.[64]

The person chosen was Timmie Jean Lindsey. Unlike the women discussed in surgical journals, she was apparently not clamouring for bigger breasts; nor was she feeling inferior as a result of the loss of volume in her postnatal breasts. Lindsey had married at fifteen years old, and was newly divorced. She met Gerow by happenstance: since she earned only £19 a week in an electronics factory, Lindsey qualified for free medical treatment. She arrived at the Jefferson Davis Hospital in Houston, Texas, with an altogether different problem. Impulsively, and persuaded by a man she was dating, Lindsey had a tattoo of a rose inked onto her breast. She regretted it almost immediately, and wanted it removed. Gerow agreed to remove the tattoo with 'dermabrasion', a procedure that takes away the top layers of the skin.

Gerow also offered to improve the physical appearance of Lindsey's chest, to insert an implant that would reduce the sagging and increase volume. Lindsey says that she had never felt self-conscious about her breasts. 'I told them I'd rather have my ears fixed than have new breasts', since she'd always been self-conscious about the way they 'stuck out'.[65] In the end, Lindsey agreed to have the implants inserted into her chest if Gerow pinned back her ears at the same time. Both surgeries were carried out in the spring of 1962. Gerow increased Lindsey's bust measurement from a B to a C cup with the use of two rubber envelope-sacs, shaped like teardrops, that were filled with viscous silicone-gel.

'When I came round from the anaesthetic, it felt like an elephant was sitting on my chest', Lindsey subsequently reported. When the bandages were removed, Gerow was pleased with the results. Although Lindsey was not particularly interested in examining her breasts in the mirror, figuring the implants were 'out of sight and out of mind', Gerow and his team were intrigued: 'All the young doctors were standing around to look at "the masterpiece," ' she said retrospectively, while acknowledging the years of 'pain and misery' they had brought her.[66] Lindsey's implants were viewed as a medical triumph. Their apparently smooth appearance and feel, and the widely reported success of the operation, led to a tremendous boom to the Dow Corning Corporation, and to Cronin and Gerow, who sold their rights to the manufacturing company in exchange for royalties. By 1970 Dow Corning had sold 50,000 implants and the future of breast augmentation seemed secure.[67]

Since their first incarnation, however, silicone implants were plagued by controversy. Some studies claim that for every woman happy with the outcome there has been another who has experienced deteriorating health, especially if the implant has torn and leaked, spilling silicone into her chest.[68] Despite a plethora of anecdotal information, however, there has been no proven link between leaked silicone and physical illness. Lindsey had periodic consultations with Gerow where he advised, even when she complained of pain and weakness, that silicone could not be to blame. This remains the official response given to women who believe they have been poisoned with silicone. After all, silicone is used in a range of medical devices including artificial valves, catheters, and as the lubricant in syringes.[69]

Nevertheless, debates over the safety and efficacy of silicone continued through the 1960s, 1970s, and 1980s, reaching its peak in the 1990s with a flood of epidemiological studies on human populations.[70] Cosmetic surgery remained unregulated, and manufacturers like Dow Corning focused their attention on developing implants that were more and more 'life-like'. Polyurethane foam was used to hold the silicone gel, a process that helped prevent capsular contracture, or the hardening of the collagen fibres in the breast as an immune response to the presence of a foreign object (i.e. the implant). The foam coating helped prevent capsular contracture, but an unfortunate side effect was that the foam could disintegrate within the body and be difficult to remove.[71]

In the US, amidst calls for its involvement in the development of technological devices of all kinds, the Food and Drug Administration (FDA) enacted the Medical Devices Amendment in 1976.[72] This meant that the FDA could approve the safety and effectiveness data of new medical devices.[73] Because breast implants were already in use they were 'grandfathered in', meaning they could remain in use without any safety evidence being required.[74] Just over a decade later, however, the regulatory landscape shifted.[75] A number of important developments lay behind the change. First, there was intensified debate in the 1980s and 1990s over the safety of silicone implants. Medical articles citing anecdotal evidence from the US and Japan began emphasizing a link between the use of siliconc and connective tissue disease.[76] A rising tide of lawsuits were issued against manufacturers by women who claimed their health had suffered after implant rupture.[77] This litigation took place within the context of a flurry of mass torts in US federal courts, many of which were dealing with what one legal scholar has termed 'large scale technological disasters'.[78]

The national and international media responded loudly, investigating the plight of individual women and criticizing the perceived negligence of surgeons and manufacturers.[79] The most famous example was an episode of the CBS television show *Face to Face with Connie Chung*, broadcast in 1990 and featuring the real life experiences of women who claimed to have autoimmune diseases as a result of silicone implants.[80] The programme alleged that unsuspecting women had had dangerous, untested devices foisted on them by an unthinking surgical profession. The FDA was also blamed for permitting hazardous devices to be sold.[81]

Presumably as a result of public pressure, the FDA shifted its position on implants in early 1991. Manufacturers like Dow Corning were now required to provide evidence of their safety. When the evidence proved inadequate, an FDA advisory panel met to discuss how to move forward. Dow Corning was then on the losing end of a lawsuit worth over $7 million from a woman who claimed her implants had caused connected tissue disease.[82] It was also charged that Dow Corning withheld internal concerns about the safety of breast implants.[83] Bernard M. Patten, a neurologist and one time colleague of Gerow and Cronin, published a series of articles in which he related breast implants to neurological and immune problems.[84] Although his attempts to expose the dangers of silicone were downplayed, he claimed that as early as 1954 an in-house study by Dow Corning showed silicone could be linked to toxicity. It seems that not all the laboratory dogs that were used in the testing were 'fine'. In some cases, silicone had migrated to the major organs, ending with the death of one animal and the development of serious inflammatory disease in three others.[85] Patten was adamant that Dow Corning was aware that silicone caused both inflammation and autoimmune diseases. Though the company denied this, it acknowledged the possibility that implants might rupture.[86]A full-scale reputational battle was then played out between Dow Corning and the media as well as in the courts.[87]

In the context of rising consumer and media pressure the FDA Commissioner David Kessler issued an immediate moratorium on silicone implants. He explained the decision in an article for the *New England Journal of Medicine*, stating that too much was unknown about the possibility of rupture, and about the impact of the chemicals that could leak into the body:

> The link, if any, between these implants and immune-related disorders and other systemic diseases is also unknown. Serious questions remain about the ability of manufacturers to produce the device reliably and under strict quality controls. Until these questions are answered, the FDA cannot legally approve the general use of breast implants filled with silicone gel.[88]

Only women whose need for them was 'most urgent' were to be allowed silicone breast implants. That included women awaiting cosmetic surgery as a result of mastectomies, for which implantation was considered part of the medical treatment, or those who had experienced rupture of existing devices. In the case of cancer patients and others who needed implants for reconstruction purposes, the FDA judged that the risk–benefit ratio ruled in favour of the use of silicone implants.[89] 'Certainly as a society,' Kessler concluded, 'we are far from according cosmetic interventions the same importance as a matter of public health that we accord to cancer treatments.'[90] The moratorium on silicone implants, Kessler was careful to point out, was not because implants had been found dangerous, but because they had not been proved safe: a muddy kind of logic that did little to pacify opponents of implants. Litigation continued, and though breast implant manufacturers continued to deny any scientific evidence that silicone breast implants caused autoimmune diseases, the largest class action settlement against manufacturers was finalized in 1994. The biggest contributor to that settlement was Dow Corning. The company subsequently, amidst a further 20,000 pending lawsuits, filed for bankruptcy.[91]

Less than a decade later silicone implants were back on the market. Since the ban numerous studies were undertaken—directed by the FDA as well as individual interest groups—to establish whether there were scientific links between silicone implants and any 'established or atypical connective tissue disorder' and illnesses that included not only 'arthralgias, lymphadenopathy, myalgias, sicca symptoms, skin changes, and stiffness', but also 'cancer, definite or atypical connective tissue disease, adverse offspring effects, or neurologic disease'.[92] In 1997 the IOM carried out a comprehensive evaluation of the evidence for the association of breast implants—silicone and saline—with human health complaints.[93] When its report was published in 1999, it concluded that of the approximately 1.8 million US women with breast implants, about seventy per cent were elective augmentation rather than reconstruction after disease. It noted, too, than a further 10 million people in the U.S. had some kind of implant, from a finger joint to a pacemaker, which was made of silicone. It acknowledged that studies of the toxicology of silicone ought to have been more long term, but that the overall findings were not harmful in the usual quantities and exposure. There was no identified proof of a connection between ruptured implants and any kind of systemic condition.[94] More specifically, the charge that there was a unique condition associated with breast implant sufferers was entirely rejected:

> Evidence for this proposed disease rests on case reports and is insufficient or flawed. The disease definition includes, as a precondition, the presence of silicone breast implants, so it cannot be studied as an independent health problem. The committee finds that the diagnosis of this condition could depend on the presence of a number of symptoms that are nonspecific and common in the general population. Thus, there does not appear to be even suggestive evidence for the existence of a novel syndrome in women with breast implants. In fact, epidemiological evidence suggests that there is no novel syndrome.[95]

There was one caveat relating to 'local problems' associated with silicone migration, and an acknowledgement that long-term quantitative studies were lacking. The only confirmed potential side effects were the local hazards linked with breast implant surgery: asymmetry, pain, atrophy of the skin, calcification and hardening of the implants, the appearance of chest wall deformity, delayed wound healing, haematoma (a collection of blood near the surgical site), damage to tissue or implant, toxic shock syndrome, inflammation, irritation, and infection that can lead to necrosis or the death of skin tissue round the implant.

These hazards have not been considered significant enough to keep silicone off the market. The complaints of many thousands of women were dismissed because they were not scientifically verifiable, or 'not consistently associated with objectively physical signs or laboratory abnormalities'. The subjective experiences listed by women were associated with conditions like chronic fatigue syndrome and fibromyalgia, both common to the wider female population (especially in 'young to middle aged women'), and also associated with 'concurrent psychiatric disorders'.[96] In short, it was now possible to argue that just as women sought cosmetic surgery because of psychiatric problems, those same psychiatric problems would lead them to be unhappy with the outcome and perhaps even to imagine physical symptoms.

In a surprising volte-face, there was a dramatic increase in the number of cosmetic breast implants after the moratorium was lifted. Over 132,000 women received silicone breast implants in 1998 alone. Moreover, increasing numbers of physicians who were not trained as surgeons started to perform cosmetic surgery.[97] Though the FDA requested better long-term tracking of implant recipients, the responsibility for the outcome of surgery was effectively pushed onto the consumer. 'Breast implants are not lifetime devices', the FDA warns. 'The longer a woman has them, the more likely she is to have complications and need to have the implants removed or replaced. Women with breast implants will need to monitor their breasts for the rest of their lives.'[98]

PIP AND DÉJÀ VU

Thus far I have dealt principally with the North American context, since that is where breast implants were first popularized. Yet in the 2000s a remarkably similar debacle over the ethics and safety of silicone breast implants took place in the UK, throughout Europe and in parts of South America. There are parallels, too, with the US in the lack of regulatory framework, and an attitude towards cosmetic surgery that promotes consumer choice, but that also imputes psychological problems to women who associate ruptured implants with other health concerns. An added nuance in the UK is the moralizing judgement associated with aesthetic as opposed to reconstructive breast implants because of the involvement of the NHS in having to correct the work of private clinics.[99]

Debbie Lewis, a hairdresser from Buckinghamshire, underwent breast augmentation surgery in 2004. Like Lindsey she had recently separated from her husband and she decided to alter her physical appearance to make herself feel better. Also like Lindsey we hear her story only as a result of media interest. Lewis' stated decision to have breast implants was psychological. She wanted to 'treat' herself after a difficult time in her life, seeking the 'pleasure' and 'self-confidence' that large breasts promised. The surgery cost Lewis £4,000 and within a year she had the implants changed twice, enduring the pain and discomfort of the surgery and its after-effects because of problems with the implant. First, Lewis experienced capsular contraction. Then she discovered that one of her implants had ruptured and leaked its contents into a lymph node. After an operation lasting an hour and a half, in which her swollen lymph nodes were removed along with the implant, Lewis' cosmetic surgeon reported that 'the implant shell looked like a thin beach ball. It was not a good quality product'.[100]

Poly Implant Prothèse (PIP) produced the implant that was removed from Lewis' body, at a cost of an additional £6,000. PIP is the French company at the centre of the latest breast implant scandal. Its products were banned throughout Europe from 2010, amidst complaints of high levels of rupture and degradation. Lewis was one of 40,000 women who received PIP implants that were later discovered to be made not with medically approved silicone, but unauthorized industrial grade and agricultural silicone—the kind used to stuff mattresses—as well as the chemicals Baysilone, Silopren, and Rhodorsil that were used as fuel additives and in the manufacture of industrial rubber tubing.[101] In December 2013 the founder of the PIP distribution company, Jean-Claude Mas, was found guilty of aggravated fraud and sentenced to four years in prison by a court in Marseilles.[102]

The PIP incident prompted a UK review into whether the cosmetic surgery industry needed to be regulated, and the government eventually agreed to remove PIP implants from affected women on the NHS. The case was of concern to more than 400,000 women throughout the world, not only in Britain but also in Brazil, Venezuela, Argentina, Italy, Germany, and Spain. From 2009 an abnormally high level of ruptures began to be reported by women like Debbie Lewis. Another Debbie, this time Debbie Davies from Liverpool, reported that one of her PIP implants had ruptured: 'so I have silicone going around my body. It is in my lymph nodes and I have lumps of it on my neck and in my chest. They are literally killing me.'[103] There is no scientific 'proof' that silicone can cause illness, as has been discussed above. This was not reassuring for Davies, especially since the implant did not contain medical grade silicone, and because recipients of PIP implants have been advised to have them removed in case of rupture. Albeit not on the US scale, class action suits are being pursued by thousands of patients against clinics and surgeons that supplied PIP implants.

How did this crisis happen? In the UK breast implants are regulated under a European Union medical device directive. Until the 1990s each of the European Union (EU) countries had its own approach to evaluating medical devices, though a Conformité Européenne (CE) mark given in any one country meant that the device could be sold throughout the EU. Device approval in the UK is approved within each member state; in the UK it is overseen by the Medical and Healthcare products Regulatory Authority (MHRA). There are limits on the system of regulation; the evidence on safety and efficacy of new devices and procedures is 'variable', even 'poor'.[104] In the case of PIP implants, the content of the implants was changed *after* the original CE mark was awarded. This is only the tip of the regulatory iceberg: silicone implants used to enhance buttocks and even calves do not need a CE mark, because they are not considered to have a medical purpose.[105]

The MHRA released a Medical Device Alert concerning PIP implants in 2010, the same year the implants were taken off the market. But France's medical safety watchdog had been aware of increasing risks from 2006, the year that UK surgeons had been reporting an overly high rupture rate.

British surgeons continued to use PIP despite their concerns. Had there been any self-regulation earlier, or any governmental intervention in the UK, thousands of women would have been spared unsafe implants.[106] Like the American situation, the PIP scandal arguably occurred because the technologies involved were not appropriately monitored, and because governments were unwilling to get involved in the complex rights and responsibilities around cosmetic surgery.[107]

In 2010 the Department of Health (DOH) estimated that between 40,000 and 80,000 UK women had received PIP implants, ninety per cent of which were purely cosmetic.[108] So far, a range of tests to evaluate human hazard have proved inconclusive.[109] Although PIP implants were acknowledged to be substandard with an increased rate of rupture, DOH found no evidence of any increased clinical risk.[110] This claim is in direct opposition to the thousands of women who identify that their health has suffered as a result.

More than ninety-five per cent of PIP implants were fitted by private and unregulated clinics. Those clinics have since been criticized for cut-price deals, giving unrealistic expectations to patients, and for aggressive selling practices that include time-limited deals, financial inducements, packages (like 'buy one get one free' or mother/daughter deals) and offering cosmetic procedures as competition prizes.[111] A largely unregulated cosmetic surgery industry has emerged, where unsubstantiated promises are made that a simple operation will mean a better life. The Harley Medical Group, for instance, one of the leading private clinics, advertises 'popular' procedures that suggests patients will receive 'the body confidence and boost in self-esteem that [they] deserve'.[112]

The pressure on women to have 'perfect' breasts comes from a range of sources, so how can we possibly know what perfection looks like? Thankfully, a cosmetic surgeon revealed all to *The Sun* newspaper in 2011, creating a 'scientific formula' for the perfect degree of perkiness (see Fig. 2).[113] The proportion of breast that should sit above the 'meridian' (where the

Fig. 2. A scientific formula for the perfect breast, created by a cosmetic surgeon for *The Sun* newspaper in 2011.

nipple with an upward angle of twenty per cent should be) is forty-five per cent. Fifty-five per cent of the breast tissue should fall below that magical line to give just the right amount of bounce. These measurements would give men assurance that their sexual partners were ripe and ready for reproduction. In the same year six scientists from the Victoria University of Wellington in New Zealand conducted a questionnaire to find out what men in New Zealand, Papa New Guinea, and Samoa thought about women's breasts.[114] Women were not invited to respond. Results indicated men from New Zealand preferred symmetrical breasts, but those from Papua New Guinea cared less about symmetry than that the breasts had large areolae. Men from Samoa preferred breasts with darkly pigmented areolae, but men from New Zealand preferred them to be lighter. Cultural differences matter in the definition of perfection, then, which must be why the human race has not died out, with so many imperfect breasts bouncing (or hanging) around.

In 2013, after a series of toing and froing about whether the NHS was morally obliged to remove implants that had been inserted privately, a judgement was made in favour of removal. Health systems in both France and Wales agreed to fund removal of PIP implants and their replacement with 'safe' alternatives. However, at the time of writing, the NHS in England will only fund the explantation or removal of the implants; except in exceptional circumstances, any replacements are made at the patients' own expense.[115] This situation raises ethical dilemmas about the moral obligation of the NHS, and the problematic interrelation of the private and national health sectors. Indeed, the whole PIP debacle raises several contentious questions about governmental responsibility and the ethics of healthcare. How did PIP sell thousands of implants worldwide for nearly a decade before its implants were found to be faulty? There has been no consistent record kept of women who received implants, nor of the impact of those implants on the psychological and physical health of the women concerned—despite the fact that breast implants are increasingly associated with mental health problems. And the subjective medical experiences of thousands of women seem to have been dismissed out of hand.

In 2013 Sir Bruce Keogh, Medical Director of the NHS, published a Review of the Regulation of Cosmetic Interventions that explored the condition of the UK cosmetic industry[116] Though the Review was prompted by the PIP case, 'which exposed woeful lapses in product quality, after care and record keeping', it was concerned with cosmetic interventions across the sector, and revealed 'widespread use of misleading advertising, inappropriate marketing and unsafe practices'. This judgement applied not only to cosmetic interventions—which were worth £2.3 billion in 2010 alone—but also to non-surgical interventions like dermal fillers and Botox. Non-surgical interventions can lead to scarring, tissue damage, blindness and facial disfigurement, and yet consumers have 'no more protection and redress than someone buying a ballpoint pen or a toothbrush'.[117]

Despite the findings of Keogh's Review, private clinics are largely unregulated. Their advertising takes it for granted that women seeking breast augmentation suffer from low self-esteem because of the way their bodies look or that they have problems with 'body confidence'. They might just want 'better shaped breasts that are in proportion with the rest of her body' as one cosmetic surgeon puts it on his website, or to 'restore the natural fullness to their breasts and feel confident in their bodies again' in the words of the Harley Medical Group—especially if they have 'aged' or have breastfed their children and endure drooping, sagging breasts.[118] For a 'no obligation' consultation with a specialist, those women could be well on the way to recovery and towards a better, younger, more attractive version of themselves. And if a woman can 'fix' herself this easily, this painlessly, this conveniently, why not turn to breast surgery to improve her life, just as she might turn to a new moisturizer or hair conditioner?[119]

Cosmetic surgery adverts on billboards, at railway and underground stations, and in magazines reassure potential patients that these are everyday operations performed by objective, highly skilled, and experienced professionals. Surgeons at the Harley Medical Group are said to 'perform hundreds of breast surgery operations every year'.[120] Surgical expertise is stressed above any social or psychological implications of the procedure. It is extraordinary that serious medical procedures involving general anaesthetic and its accompanying risks have been framed as lifestyle choices.[121] Keogh's Review acknowledged that procedures should not be sold as a 'commodity', and also that the current regulatory framework did not ensure patient safety or best practice.[122] The British Association of Aesthetic Plastic Surgeons (BAAPS) has also argued that people have been dangerously misinformed about the risks and likely outcomes of surgery by salespeople working to targets, rather than qualified surgeons.[123]

It was not until 2012 that medical revalidation was introduced to monitor surgeons' skills and abilities. Prior to that over ninety per cent of people simply presumed the surgeon was 'fully qualified' in whatever procedure she—or usually he—was going to undertake.[124] As part of a regulatory overhaul Keogh recommended that the Royal College of Surgeons of England (RCS) should establish a Cosmetic Surgery Interspeciality Committee to regulate and monitor the UK cosmetic surgery industry, set standards for training and practice, ensure competence of surgeons, establish a clinical audit database, and work with the Parliamentary and Health Service Ombudsman (PHSO) on dispute resolution.

Despite Keogh's comprehensive report, there has been a lack of impetus on the part of the government to address its recommendations. This is rather different to the American and even the French governmental response, where reactions to scandals have included a tightening up of surgical and non-surgical cosmetic medicine.[125] This apparent UK resistance to monitoring cosmetic surgery is not new. Melanie Latham has shown how breast implant risks were highlighted to the UK government over many years prior to the PIP scandal. A National Care Standards Commission (NCSC) Report published in June 2003 catalogued the ways cosmetic surgery clinics in Central London were not adhering to the minimum standards required by law.[126] Better self-assessment and self-regulation were recommended rather than governmental intervention. Even earlier, fears about leaking silicone breast implants were rife in the 1990s, prompting a governmental enquiry into the relationship between implants and connective tissue disease.[127] This enquiry was set up as a response to the FDA's moratorium on breast implants, though the Report to the Chief Medical Officer ruled that there should be no change in practice in the UK based on America's experience. Its single concession was to introduce the National Breast Implant Registry to collect data on those who had undergone breast implant surgery. Mr Brian Morgan, a Consultant Plastic Surgeon at University College Hospital, worked closely with the government to establish the register from 1992, but it was closed in 2006 due to poor take-up and lack of funding.[128]

In June 1994 Ann Clwyd, MP introduced an unsuccessful cosmetic surgery bill to establish minimum standards of training and practice. In 2012 she introduced a further private member's bill, which got no further than its first reading.[129] Whether or not things will be different in the wake of the PIP scandal remains to be seen. There has been an increase in demand for implants manufactured in the UK; for instance the Glasgow-based company Nagor reported a thirty per cent increase in sales in 2012.[130] Choose your implant as well as your surgeon carefully, women are now being told as good consumers. Unfortunately the social and psychological reasons why women might be choosing implants are still being overlooked, and the emphasis is placed on surgeons to self-regulate. The RCS has published its own statement of standards for cosmetic practice. The guidelines suggest individual surgeons should:

> Discuss relevant psychological issues (including any psychiatric history) with the patient to establish the nature of their body image concerns and their reasons for seeking treatment. They should not at any point imply that treatment would improve a patient's psychological wellbeing.[131]

Moreover, surgeons should consider whether a psychological assessment is necessary before operating, as well as ensuring that any advertising is honest. These guidelines represent an apparent attempt by the RCS to protect its professional reputation by distancing itself from poor standards, false advertising, and bad practice. Yet as Professor Norman Williams, former President of the Royal College of Surgeons, acknowledged in 2013, the RCS is not a regulator, nor a legislator. Without governmental intervention it is unclear how far cosmetic surgery standards can or will be improved.[132]

FROM PATIENT TO CONSUMER: LIBERATION OR LIABILITY?

Timmie Jean Lindsey, the world's first silicone breast implant recipient, is now in her eighties and still working nights at a care home. Many of the women that were subsequently enlisted in Cronin and Gerow's trial—including relatives and friends of Lindsey herself—apparently became ill after the procedure. Some died, allegedly because of the leaking silicone, though as this chapter has shown, such claims have no scientific justification.[133] Lindsey has come to terms with her own implants, despite the pain and discomfort they have caused: 'You can feel the prostheses if you press, but I have enough breast tissue so it really isn't that noticeable. And they've aged, like I have. They haven't stayed straight up and perky; they've drooped, too. They feel like part of me now.'[134]

Cosmetic surgery is one of the largest growing medical specialisms, with 'lunchtime' treatments for non-surgical interventions (Botox to stop lines from forming, fillers to 'fill them in') and surgical interventions packaged as self-improvement in a way that a century ago would have been unconvincing. Small or saggy breasts, bent penises, skin-stretched bellies and too long labial lips have been redefined as clinical anomalies rather than facts of life. Living with such 'deformities' (previously regarded as the luck of the draw) has been seen to cause depression and a lack of confidence.

The PIP breast implant scandal has not prevented women from calculating that the desired outcome—a perkier, tighter, sexier appearance, which might lead to a better relationship, job, or quality of life—is worth the financial outlay, the discomfort, and the risks involved. What does it say about our discomfort and dis-ease as a nation that we take these risks? And how has it become a personal preference, or a matter of women 'choosing wisely' if seeking breast implants in the private sector, when their decisions are, according to the Keogh Review often based on misunderstanding about the risks and guarantees involved?[135] These are challenging questions, and there are many more.

So far I have focused on the construction of need as identified by the medical profession, and there are important reasons why this is so. The story of why women might seek breast implants, now or in the 1950s, is a long and complex one. It is too simplistic to see women as passive recipients of cultural expectations about the perfect body, and yet we know that the steady drip of negative images and self-hatred can have devastating effects on a woman's psyche.[136] We also know that body image is a major factor in women's choice to have cosmetic surgery and that it is on women's bodies that over ninety per cent of all British cosmetic surgery takes place.[137]

Traditionally, the story of silicone breast implants is told from a number of distinct, polarized perspectives—from the viewpoints of the surgeons, keen to deliver on women's expectations; of the manufacturers that develop and deliver products; or of women themselves, often those whose operations have been unsuccessful.[138] We are used to thinking of female patients as victims, that their bodies are colonized, their minds conned into thinking that breast implants will lead to contentment.[139] But within feminist scholarship, the popularity of breast implants is problematic, especially among women who do not self-identify as intellectually downtrodden or uneducated.[140] There has been a remarkable take-up of cosmetic surgery among educated and feminist women.[141] Despite the fact that their surgeons are predominantly white, male, and middle class, we have seen some feminists embracing the potential for transformation as a positive act: cosmetic surgery as a technology by which women can forge new identities. Of course those new identities ultimately project a predominantly white, Western notion of attractiveness.[142] Those who genuinely try to transform themselves into radically different forms–into tigers, say, or lizards–are few and far between.[143]

It is hard not to see breast implants as symptomatic of the cultural and aesthetic demands of patriarchy, and of the impossible ideals of youth and beauty to which women are routinely subjected.[144] A disproportionately high number of implant patients report distress over their appearance. Many women who undergo breast augmentation also suffer low self-esteem, depression, and body dysmorphia.[145] Women seeking breast implants are believed to be three times more likely to commit suicide than those who do not.[146] Breast implant patients are more frequently linked with depression and suicide than any other cosmetic surgery patients.[147] Susie Orbach has suggested that the self-hatred or criticism of an individual towards a specific body part reflects some latent and unresolved childhood conflict that is symbolized by that body part.[148] Whatever the case, like operations to alter 'unsightly' labia or to replace the hymen, breast augmentation surgery seems to strike at the heart of femininity. It suggests a degree of self-regulation applied to the female body that is startling in its extremity.

Technological and surgical developments do not take place within a vacuum. Nor do ideas about psychological need or scientific progress. The treatment of women as consumers and patients by the surgical profession needs to be addressed, especially since many breast implant recipients feel ignored or neglected by those professionals they have consulted. In this context the debates on scientific 'proof' are arguably less important than the fact that women's physical and psychological suffering is hystericized in the twenty-first century just as it was in the nineteenth.[149] There are similarities in the use of power and authority around breast implants and other health concerns that affect women, including the use of oestrogen in cases of osteoporosis, the diagnosis and treatment of fibromyalgia as a disease entity and the treatment of premenstrual syndrome (PMS).[150] Few cosmetic surgeons today are women, and the profession remains male-dominated.[151]

The overwhelming evidence is that bigger breasts rarely, if ever, bring a better life, more luck in love or a more successful lifestyle. In fact they are more likely to bring additional pain and suffering. There is no 'quick fix' for most of the hopes and fears that take women to the cosmetic surgeon. And yet many of us still dream that a 'better', more youthful body will bring these things. We cannot disassociate breast implant surgery from the aspirations of the culture in which it takes place, nor from the professional structures that perpetuate and maintain it. And we cannot ignore the profiteering that takes place on and around female bodies, or the lack of a regulatory framework that might protect women as health consumers.

Breasts are not unchanging, immutable signifiers of womanhood: they evolve from puberty, through sexual maturity, as well as through childbirth and breastfeeding or illness or weight loss. Breasts appear differently depending on what clothes are worn, what enhancements are used, and what artwork might adorn them. In 2011 the photographer Laura Dodsworth interviewed 100 women between the ages of nineteen and 101 to ask them how they felt about their breasts, producing a photo gallery that was somewhat removed from idealized surgery adverts or the airbrushed pages of *The Sun*.[152] In an interview with *The Guardian*, Dodsworth suggests that 'while breasts are interesting in themselves, they are also catalysts for discussing relationships, body image and ageing'. Physically and psychically, breasts are marked with women's experiences as human beings. They are part of what the feminist philosopher Susan Bordo, in another context, calls the 'complex crystallization of culture'; a lens through which we can see both the expectations of cultural ideals and the ways individual women respond to those ideals as embodied human beings.[153] Breasts are far more than pillows, then, whatever Anatole France might think. For good or ill they are stamped passports of women's interactions with the world.

Women's bodies have long been battlegrounds for debates about how they should look. In the pursuit of perfection the boundaries between subjective experience and cultural expectation can be blurred, even when the decision to go under the knife is celebrated as a post-feminist or liberatory ideal. It is not only breasts that have been measured and found lacking against an externally imposed norm; nor the only time that surgery has seemed a viable and even feminist solution to women's perceived inadequacies. Female genitalia are also subject to the same kind of social and medical scrutiny, as the following chapter demonstrates.

3

'Country Matters'

The Language and Politics of Female Genitalia

HAMLET	Lady, shall I lie in your lap?
OPHELIA	No, my lord.
HAMLET	I mean, my head upon your lap?
OPHELIA	Ay, my lord.
HAMLET	Do you think I meant country matters?
OPHELIA	I think nothing, my lord.
HAMLET	That's a fair thought to lie between maids' legs.
OPHELIA	What is, my lord?
HAMLET	No thing.

(*Hamlet*, III. ii. 107–115)

Hamlet's exchange with Ophelia is a ribald one. Elizabethan audiences would have understood the sexual puns in Hamlet's words—lie (to have sex with), head (penis or oral sex), country matters (allusion to 'cunt') and 'no thing' (a slang term for a vagina which reveals its definition by absence, i.e. no penis).[1] Much has been written about Shakespeare's frequent references to sex and sex organs, and there are perhaps 700 puns about sex in the plays alone.[2] There is also a considerable body of analysis on the troubling nature of female sexuality and desire in Shakespeare's work. Women arguably feature most often in Shakespeare principally in terms of their sex, and not as a focus for social criticism or observation but as a source of mythical power that arouses both love and loathing in their male counterparts.[3]

Historians of the so-called 'one-sex model' of gender difference argue that early modern men and women viewed sexed difference on a continuum; that the female body was essentially an imperfect version of the male. Thus some anatomists wrote about the vagina as an internal penis, the labia as the foreskin, the uterus as scrotum and the ovaries as testicles, an inversion made possible by the physical imperfection of women; unlike men they lacked the sufficient heat for the sex organs to be drawn outside of the body and to become male.[4] This is an interesting historical hypothesis, as is the suggestion that the Enlightenment saw a complete separation of male and female into our modern 'two-sex model' of binary biological difference. But it also depends on a rather simplistic interpretation of a few select anatomical texts. Not all early modern writers were wedded to the idea of the one-sex model, and there were many anatomists who viewed male and female organs as entirely different.[5] Additionally, there was a degree of animosity towards the vagina in the early modern period, as indicated by Shakespeare's prose. The vagina was a troubling space, an organ of creative power and maternity, but also of sexual manipulation and control.[6]

One of the most striking aspects of Shakespeare's references to female genitalia is the range of words used to describe them: the vagina is not merely a no thing (*Hamlet*), but a 'vallie-fountain' (Sonnet 153), a 'deep Pit' into which one might stumble and be lost (*Titus Andronicus*), a bracelet (*Cymberline*) and a 'darke and vicious place' (*King Lear*).[7] The vagina has many names but no name; it is an object of euphemistic discourse in the past as in the present, as testified by Eve Ensler's *Vagina Monologues* and its contributors, many of whom identify with what Ensler calls the '"down there" generation'.[8]

Even today, a study of dictionary definitions of 'vagina' reveals that the organ is seldom described in relation to sex (unlike the penis), but overwhelmingly in terms of its physical location. Similarly, while 'penis' is almost always described as an *organ*, the vagina is not; it is pictured as a 'canal' or a 'passage' that by implication leads somewhere else. Moreover the clitoris is often missing from those dictionaries; where it has been included that is principally in relation to the penis.[9] The possession of a vagina, or the absence of a penis, is what has historically defined female against male bodies. This binary difference is not only problematic for women (whose bodies are defined by that absence), but also for representatives of the LGBT+ movement, especially in relation to transgender identity.[10] There is ample literature to show that women have been delimited by, and reduced to their vaginas for centuries, which raises the question whether the equation of womanhood to the vagina in Ensler's monologues is just another way of reducing women to their sexed existence.[11]

Leaving behind the one-sex model, this chapter explores the ways women's sex organs have been understood in history: most notably the vagina, the vulva, the clitoris, and the hymen. Since Shakespeare's time female genitalia have been written about in a number of different ways, anatomical explorers laying claim to specific body parts much as their seafaring counterparts colonized new geographical lands. In the process the mapping of the female body became a powerful social and political metaphor for gendered discussions about the dominance of reason over passion, of civilization over nature.[12] Female genitalia are often the subject of multiple anxieties over the control of women's bodies and their reproductive function. The reasons have been theological and ideological as well as purely economic, such as primogeniture, the passing down of property through the male heir, and religious ideals of female chastity.[13]

In the twenty-first century, governments, pressure groups, feminists, educationalists, and religious organizations express concerns about female sexuality, linked both to reactions to the so-called 'sexual revolution' and the ever-present double standards of sexual morality. Control over female sexuality is embedded in broader political debates and specific human rights abuses like female genital mutilation and sexual slavery.[14]

The words used to describe women's sex organs also remain controversial, whether that is the profoundly provocative 'cunt' or deliberately demure 'lady garden'.[15] Political and linguistic anxieties reflect not only cultural concerns about women's bodies as sexually active or sexually reproductive, but also political values about women's sexual and socio-economic function. There is more, then, to 'the woman's part' than meets the eye.

PASSIVITY AND THE 'WOMAN'S PART(S)'

Female sexual organs are more physically complex than indicated in the generalized diagrams of reproduction found in school teaching; the two-dimensional sketches that highlight the fallopian tubes, the ovaries, the womb, and the vaginal passage that can stretch to accommodate a penis as well as a baby's head. The 'facts of life' are ordinarily explained in the most rudimentary biological terms.[16] The language in which reproductive sex is described is also important, because it sets up and confirms the social relationships in which it takes place. The vagina is figured as a passive recipient that is 'penetrated' by the penis, which deposits sperm that must race to the patiently waiting egg. This version of the physical process denies any active agency on the part of the uterus, most notably through orgasm, in dipping to pick up the sperm and help conception to take place. In 1999 the feminist Germaine Greer explored the idea of a 'different version of female receptivity by speaking of the vagina as if it were active, as if it sucked on the penis and emptied it out, rather than simply receiving the ejaculate'.[17]

To date there has been no scientific evidence confirming that the vagina, or the uterus (through hormones released at orgasm that cause muscle spasms), has an active role in propelling sperm up the vagina. It is highly relevant that evidence-based work is undertaken in women who are not sexually aroused. There is not, moreover, much financial incentive for research in this area.[18] Yet the evolutionary function of the female orgasm is a subject that has aroused much public and academic interest, especially since the 2005 publication of Elisabeth Lloyd's *The Case of the Female Orgasm*.[19] The briefest consideration of the history of the female orgasm makes clear that Greer's observations are not new. The association of female orgasm with pregnancy has a long history. Since the ancient Greeks a prerequisite to pregnancy has been orgasm in *both* parties.[20] Unfortunately, agency is a double-edged sword: this belief has been used to deny rape: from the advice of a seventeenth-century midwife, Jane Sharp, that 'extream hatred is the reason why women seldom or never conceive when ravished' to the Republican Senate nominee Todd Aitken's farcical distinction between rape and 'legitimate rape' in 2012.[21]

The language of female genitalia has changed. In early modern writing, the term 'womb' was sometimes used to describe both the womb and the vagina, treating them as one distinct unit. Sometimes the term 'bottom' is used for womb, and 'womb' for vagina.[22] Today there is a similar slippage of terms from one part of the female anatomy to another. When most people use the term 'vagina' they are usually not speaking precisely about the internal, muscular tube that connects the uterus to the fallopian tube and the ovaries, but its visible outer layer and opening (sometimes called the *vestibule* in medical textbooks), as well as the labia majora (outer lips), labia minora (inner lips) and the clitoris. There is seldom any mention of the vulva in diagrams of the female reproductive system. Yet the vulva has a sexual function; it is not a merely a gateway to the vagina. The external organs are filled with nerves that provide pleasure when properly stimulated. How strange then, that the vulva is ignored in discussions of reproduction, along with the clitoris and the hymen, despite the fact the former is crucial in enabling women to reach climax, as discussed below, and the latter in 'proving' virginity.

THE VULVA MONOLOGUES

Looking at the vulva became particularly fashionable at the end of the twentieth century, with its clinical reconstruction for aesthetic reasons. It has been suggested that women who request vulvar reconstruction overwhelmingly base their aesthetic ideals on their childlike rather than their adult appearance.[23] And this trend has its roots, somewhat alarmingly, in both the sexualization of young bodies and the fashion for the so-called 'Barbie doll' aesthetic found in pornography, in which women's external genitalia are dainty, pink, soft, even-textured, and hairless.[24] Labiaplasty, the reduction and cutting away of the inner labia, is one of the fastest growing forms of cosmetic surgery in the West.[25] In 2008, in reference to this troublesome cultural trend, the Brighton-based artist Jamie McCartney created *The Great Wall of Vagina*; a sculpture comprising plaster casts of 400 women's genitals to show how varied vulvas might be (see Fig. 3). The artist describes the work as 'art with a social conscience'.[26]

It seems churlish to point out that McCartney has actually created a Great Wall of Vulva, since the geographical allusion would be lost. On the one hand, the declared intent of McCartney, to celebrate the diversity of womanhood, to make women 'feel better' about their differences is laudable. And yet on the other, there is something oddly jarring about a male sculptor exhibiting female body parts so explicitly; for art or for profit, such a mass display is somehow objectifying (the very opposite of

Fig. 3. Panels from *The Great Wall of Vagina* exhibition by Jamie McCartney at the Hay Hill Gallery in London, 2012.

McCartney's claim, in fact, that the multitude somehow makes the depiction *less* so). There is, moreover, a subtle allusion in McCartney's work to the writings of the Renaissance scholar and humanist François Rabelais, though this may be accidental. In *Gargantua and Pantagruel,* Rabelais imagines a wall of vulvas protecting Paris, being 'cheaper than stone', given how much women's private parts sell for on the city streets.[27]

There are disturbing parallels between the rise of cosmetic surgery as an aesthetic choice and the incidence of female genital mutilation (FGM), also known as 'cutting' or 'female circumcision'. It could even be defined as such, if we use the definition given by the World Health Organization (WHO) as 'all procedures that involve partial or total removal of the external female genitalia, or other injury to the female genital organs for non-medical reasons'.[28] Of course, not all forms of cultural compliance involve surgical alteration. Another example where female genital appearance has developed a strong visual aesthetic is in the case of pubic hair.

TO BARE OR NOT TO BARE

Female body hair removal has become ubiquitous in the modern West, especially in the UK and North America.[29] One psychologist has called hairlessness, especially of the underarms and legs 'a major component of "femininity"; a norm that has developed in the United Stated since the early twentieth century'.[30] The fashion for shaved or waxed pudenda is more recent. In 2002 commentators began to discuss the ubiquity of the 'Brazilian', the removal of all hair from women's genital area that was the subject of considerable media attention from HBO's *Sex and the City* to popular magazines and websites.[31] How did we get to a place where the almost complete removal of public hair, a process that is painful, inconvenient, expensive, and repetitive, is not only fashionable but also the idealized 'norm', female body hair being seen, at least in most of Europe and the United States as unattractive, unfeminine, and even dirty?[32]

In the same way that small breasts became abnormal in the nineteenth century, excessive body hair was pathologized and associated with excessive masculinity, and with animalism and primitivism in the context of evolutionary theory. A series of high-profile case studies from the 1850s, from bearded ladies to human zoos to the 'hairy family of Burma', meant that excessive hairiness was medically and socially linked to a new condition—hypertrichosis—and women with superfluous hair on their bodies or faces were regarded as non-women. The following century, moreover, saw first the rise of feminism and the rejection of beauty ideals around the removal of body hair, before a backlash against that movement and the imposition of even stricter ideals of femininity and hairlessness.

Julia Pastrana toured Europe in the 1850s, exhibiting herself as the 'Bearded and hairy lady' (see Fig. 4).[33] Pastrana had a full beard and hairy limbs, and she was a subject of fascination to physicians and dermatologists as well as the general public. In 1857 she arrived in Britain from America and was known as the baboon-woman, seen as the missing link between humans and animals—especially with the publication of Charles Darwin's *Origin of Species*.[34] This evolutionary perspective marked an important shift from the early modern period, when accounts of excessively hairy people like the González family of Tenerife were associated with mythological 'wild men' and creatures of fable rather than scientific examples of racial difference.[35] Even after she died in childbirth Pastrana didn't fail to entertain: her husband-manager continued to tour with her embalmed body. As late as the 1970s her body was exhibited in Sweden, before it ended up, amidst waves of protest about the disrespectful display, in Oslo's Institute of Forensic Medicine.

Pastrana was not the only 'bearded lady' to be exhibited in Victorian Britain. The 1870s saw numerous exhibits and spectacles throughout Europe, including the 'human zoos' set up in cities like Warsaw, Hamburg, Barcelona, Paris, and London that displayed 'negro villages' and contributed

Fig. 4. Julia Pastrana: 'the bearded and hairy lady'.

to ideals of racial difference between Europeans and non-Europeans, between the familiar and the exotic.[36] Hairy women were part of this narrative of racial as well as gender difference, as they seemed to partake of both male and female characteristics. In 1877 the American physician and professor of dermatology Louis Adolphus Duhring published 'A Case of a Bearded Woman'.[37] Describing a young, physically healthy mother called 'Viola', who sported a full beard and whiskers, Duhring pondered the relationship between hairiness and masculinity, and whether a 'real woman' could have a beard. He was not alone in his speculations: between the 1870s and the 1920s, numerous dermatological reports and articles about 'hypertrichosis' or excessive hair, revealed that female patients felt traumatized by unwanted hair on their faces and bodies, as it made them less feminine, less womanly, and more like men. As the historian Kimberley Hamlin has put it, 'if the absence of hair was supposed to distinguish women from men and humans from animals, then hairy women could only be diseased'.[38]

Perhaps unsurprisingly, given that hypertrichosis seemed to afflict ordinary women and not just bearded ladies, the removal of body hair became a subject of intense discussion within the new specialism of dermatology. From the 1880s medical journals worried about the incidence of hair on women, and the subject of hair removal. Body hair had long been linked to a lack of natural womanhood: consider Banquo's confrontation with the witches in *Macbeth*: 'You should be women | And yet your beards forbid me to interpret | That you are so.'[39] By the 1870s, excessive body hair was still considered unattractive, unfeminine, and unnatural, but it was also redefined as an affliction requiring 'treatment'. Unless, of course, a woman was over the magical age of thirty-five, in which case it ceased to matter as she was past the age of her sexual allure.[40] By the early twentieth century treatment had become crucial for the psychological as well as the physical health of women. As the American dermatologist Ernest McEwen explained in 1917:

> The woman afflicted feels herself an object of repulsion to the opposite sex, and as a result, set apart from the normal members of her own sex. She realizes that she bears a stigma of the male and that she does not run true to the female type; therefore, every female instinct in her demands that the thing which marks her as different from other women be removed.[41]

As in the case of breast augmentation, medical intervention became a way to counter emotional distress as a result of bodily inadequacy. Between 1915 and 1945 magazines aimed at women, such as *Harper's Bazaar*, disseminated the ideal of hairless white female beauty, making hair removal a normal part of a woman's beauty regime.[42] Hair removal was a requirement for social and economic advancement as well as for femininity and attractiveness. At a time of high immigration and increasing urbanization, advertisements aimed at women promoted hair removal and associated it with cleanliness, science, and class mobility.[43] It was not until the 1970s, however, that pubic hair came under the same scrutiny as the rest of the body.[44]

Today pubic hair removal is the norm, at least in North America and most of Europe.[45] The 'Brazilian'—the removal of all pubic hair save for a line down the centre of the mons pubis, or the Hollywood, in which the hair is entirely removed from the genital region—has become commonplace in Western culture.[46] As noted above, this rapid transformation in pubic hair aesthetics has been driven, at least in part, by the porn industry.[47] The promotion of the naked, hairless vulva as an ideal aesthetic is problematic because it infantilizes women while apparently sexualizing young girls.[48] Moreover, the timing by which pubic hair removal became widespread may be significant.

In an article for *The Huffington Post* in 2011, Roger Friedland, Visiting Professor of Media, Culture, and Communication at New York University, convincingly argued that it was at the peak of 1970s feminism, when hair equalled strength and liberation, that there was a rise in the number of films and magazines devoted to child-like, hairless women.[49] In 1974, for instance, Larry Flynt, the American publisher of *Hustler*, which specializes in full-frontals, published *Barely Legal*, which marketed explicit images of eighteen-year olds, who often looked far younger. Movies began to focus on underage characters as sexually available beings. Thus in 1976, Jodie Foster played a twelve-year old prostitute in *Taxi Driver*; and in 1978 Brooke Shields played an underage prostitute in *Pretty Baby*. Both actresses were under the age of consent when they performed the roles.

At the same time as this 'teen fetish' became mainstream, the feminist movement was making an impact on the social and political scene. In the US the Equal Rights Amendment (1972) enshrined in law the equal treatment of men and women, while in the UK the government passed the Equal Pay Act (1970) and the Sex Discrimination Act (1975). Contemporary feminists linked social and political and sexual resistance with the growth of bodily hair; Germaine Greer noted how women's failure to remove their hair was a direct challenge to the status quo.[50] Prepubescent, virginal young women presumably provided an alternative and less challenging feminine ideal.[51]

Virginal women have always been politically and socially significant. The word 'virgin' derives from the Old French *virgine*, from the Latin *virgo*, indicating a woman who has never had sex, who is sexually 'intact' and chaste. The synonyms for virgin—untouched, unspoiled, immaculate, pristine—are indicative of the moral loading around women's sexual experiences (the 'Madonna/whore' double standard that is well known in contemporary culture). In the case of women's bodies, moreover, much debate about virginity has centred on the possession of an intact hymen.

THE 'KNOT OF VIRGINITY': THE SIGNIFICANCE OF THE HYMEN

The word 'hymen' comes from the ancient Greek for membrane: 'hymenaeus'. This term was used to describe a particular type of tissue or light film that wrapped together all vital organs and bones as well as covering the entrance to the vaginal passage.[52] Invisible from the outside but deeply important for patriarchal, religious, and social reasons, the hymen has enormous significance as a badge of virginity. The historian Kathleen Coyne Kelly has argued that it was not the only symbol of virginity in the mediaeval period and that there were many others that were equally important. But nevertheless, as she acknowledges, in its representation even more than in its anatomy, the hymen has been used to 'fix' experience to create a visible before and after in the construction of virginity as a physical and even a holy state.[53]

Images and ideas about the virgin birth and the mother of Christ, the Virgin Mary, shaped many Western ideals about virginity, motherhood, and female sexuality.[54] For Christians, the miracle of the virgin birth fulfilled the biblical prophecy: 'Behold a virgin shall be with child, and shall bring forth a son, and they shall call his name Emmanuel' (Isaiah 7:14). Throughout history the hymen or 'maidenhead' has been viewed as the literal and figurative evidence of a woman's virginity, the ideal state of womankind prior to marriage. Indeed, depictions of Mary's intact hymen lay behind artistic representations of the Virgin Mother as an un-entered garden or a pane of glass that the sun entered without breaking. Such images were popular with Northern European painters of the fifteenth century, for example Robert Campin's *Mérode Altarpiece* (*c*.1426), which depicts the Virgin near rays of sunlight, depicted as golden shafts that pass through a glass pane.[55]

The midwife Salome appears in several apocryphal Gospels, and in the Gospel of James, also known as the Protovangelium of James, written about 145 CE, which includes the infancy stories in the Gospels of Matthew and Luke as well as an account of Mary's life. This is the oldest source to assert the virginity of Mary prior to and after the birth of Jesus, and it claims that was proven by the physical presence of her hymen. The text recounts how another midwife told Salome that Mary was a virgin, to which she replied:

> As the Lord my God liveth, if I make not trial and prove her nature I will not believe that a virgin hath brought forth . . . And the midwife went in and said unto Mary: Order thyself, for there is no small contention arisen concerning thee. And Salome made trial and cried out and said: Woe unto mine iniquity and mine unbelief, because I have tempted the living God, and lo, my hand falleth away from me in fire.[56]

As a result of her lack of faith, Salome's hand is withered, and presented as such in in the *Nativity* (1420) by Robert Campin.[57] Throughout the Renaissance the hymen featured in medical and anatomy texts and courtroom discussions as a natural sign of women's virginity and lack of sexual history.[58] Midwives continued to practise 'virginity tests' by manually checking for a hymen with much the same authority that they practised pregnancy tests in cases of bastardy or illegitimacy by examining the belly and the breasts for evidence of swelling, stretch-marks, or milk.[59] It was also customary for a newly married woman's bloodied sheets to be displayed to family members and friends after the wedding night as evidence of the ruptured hymen. Ballad writers poked fun at the phenomena of 'artificial virgins', and the measures to which women went to retain the allusion of virginity, whether for social respectability or for the sanctity of the marriage bed.[60]

Compulsory virginity tests were carried out into the twenty-first century in regions like Egypt, Turkey, and Mexico, where a premium is placed on female virginity prior to marriage.[61] The presence of an intact hymen (verified by doctors, midwives, or a community's elders), or of bleeding during intercourse, is used as physical proof of a woman's purity.[62] The burden of material proof has been commercialized. Today it is possible on the Internet to buy artificial hymens that 'bleed' after sex so that the necessary stain is apparent on the bed-sheets: 'Insert the Artificial Hymen into your vagina carefully,' the manufacturers say, 'when your lover penetrates, it will ooze out a liquid that appears like blood, not too much but just the right amount. Add in a few moans and groans and you will pass through undetectable!'[63] Some women are even signing up for hymen reconstructive surgery, a step beyond the 'designer vagina' that tidies up women's inner labia for cosmetic reasons.[64]

Many researchers and feminists argue that the hymen is a social and cultural myth, based on deeply rooted stereotypes about women's sexual status.[65] In any case, the importance of female virginity came before the identification of the hymen as a representative organ, for not all medical writers agreed that the hymen existed. In the fourth century Christian doctrine maintained the vagina had a natural seal or hymen that was evidence of and associated with virginity. This belief was confirmed by anatomical dissection both in the Arab world, by Ibn-Sīnā, and in the West by Andreas Vesalius.[66] The Italian surgeon Guilelmus de Saliceto agreed, describing the hymen as a 'knot of virginity' found at the entrance to the vulva.[67] Yet some anatomists saw membranous coverage of the vulva as an anomaly and potentially harmful. Helkiah Crooke, the Court physician to King James I, noted these disagreements in his treatise *Mikrokosmographia: A Description of the Body of Man.*[68] In a section entitled 'The Membrane called hymen and the marks of virginitie', Crooke wrote that 'almost all physicians think there is a certain Membrane, sometimes in the middest of the neck of the wombe . . . this membrane they say is perforated in the middest to give way to their [menstrual] courses and is broken or torn in the first accompanying with men'. But he also acknowledged that some saw the existence of the hymen as a 'meere fable'.[69] The French surgeon Paré went further, asserting that the hymen did not exist; it was a 'primitive myth, unworthy of a civilized nation like France' pedalled by the uneducated and by female midwives with excessive faith in tradition.[70]

By contrast the French physician Jacques Moreau de la Sarthe maintained that the hymen was just so thin that many anatomists missed it. It might, he added, also be broken by a woman rubbing herself too vigorously or by the activities of lesbians.[71] Disputes over the role and existence of the hymen continued into the eighteenth century. As the French naturalist Georges-Louis Leclerc, Comte de Buffon, put it, there was 'nothing less certain than these imagined signs of a body's virginity'. Men had imagined a concrete symbol for virginity, he maintained, because it was socially important. The Swiss anatomist Albrecht von Haller vehemently disagreed, arguing that nature had created the hymen to give women a moral rather than a physical purpose.[72] These debates are instructive in highlighting the social and moral preoccupations that have historically surrounded women's genitalia. As the historian of science Londa Schiebinger has argued, the detailed examination of women's sex organs preoccupied anatomists in their accounts of how female humans and female animals differed; along with the position of the vagina and the urethra, which were unique to humans, animals did not have a hymen. By contrast, anatomists compared men to animals not in relation to their reproductive system, but to such loftier issues as the possession of reason, culture, speech, and a soul.[73]

The breaking of the hymen, however symbolic it might be as a physical act, has received little attention in psychoanalysis. Some researchers are convinced that the unconscious resonance of the loss of virginity and the hymen deserve further exploration.[74] Naomi Wolf's 2012 book *Vagina: A New Biography* addresses this relationship between the vagina and the psyche as though this is a scientific revelation, but of course the brain and the vulva, the nervous system and the entire female anatomy has long been understood to be intricately connected.[75] For the psychoanalyst Sigmund Freud, the first penetration of the vagina resulted in the destruction of the hymen and provided evidence of the psychological domination of woman by man. Indeed, Freudian psychoanalysis linked 'frigidity' or female disinterest in sex, to the fear of this 'defloration'.[76] This 'deflowering', this physical gateway between one state of being (virgin) and another of non-being (non-virgin) has become part of the way we talk about the psycho-social importance of sex.

Freud's understanding of the hymen must be placed within the more general context of his vision of the vagina as psychologically threatening: 'Probably no male human being is spared the terrifying shock of threatened castration at the sight of the female genitals,' Freud wrote in his paper 'Fetishism' in 1927.[77] A few years earlier he had compared women's genitals to the head of Medusa, another monstrous female creature of mythology.[78] In his demonization of the vagina as a source of anxiety, Freud tapped into earlier narratives around the *vagina dentata*, the toothed vagina that offered a real rather than a symbolic castration hazard.[79] In these narratives the vagina is frightening, all-consuming, destabilizing, and a threat to the male self.

Freud was also suspicious of the idea that women might enjoy orgasms without penetration. In his 1905 *Essays*, Freud stated that vaginal orgasms were the only true and mature orgasms, the clitoral orgasm being an adolescent phenomenon. No evidence was given for Freud's assertion, yet it became commonplace to think of vaginal orgasm, achieved through penetration, as the only 'true' orgasm—a hierarchy that affirms a woman's reproductive function as well as her dependence on the penis. Thus *The Sexually Adequate Female* (1953) advised that 'whenever a woman is incapable of achieving an orgasm via coitus, provided the husband is an adequate partner, and prefers clitoral stimulation to any other form of sexual activity, she can be regarded as suffering from frigidity and requires sexual assistance'.[80] Freud based his theories not on female anatomy, but on his presumption of women being inferior to men, and their sexual experiences as dependent on male penetration of the vagina. This marked a remarkable neglect of the clitoris, the only organ of the body (male or female) whose sole function is to give pleasure.[81]

LOCATING THE CLITORIS

Like the hymen, the clitoris has been the subject of considerable speculation and debate about its location, its function, its status, and even its existence.[82] Physiologically, the clitoris is, like the tip of the penis, covered in nerve endings: the clitoris has more than 8,000. Unlike the penis, the function of which is urinary and reproductive, the clitoris has no role except to facilitate orgasm. Yet the development of the clitoris is in many ways related to the development of the penis. That is not to say, like Freud, that it is a 'truncated' or somehow incomplete penis. Both male and female foetuses develop sexual organs from the genital tubercle. If the foetus is male, the tubercle becomes the penis. If female, it initially grows into two separate corposa carnova, which combine into the clitoris as the foetus develops.[83]

The clitoris was first described in anatomical texts in the sixteenth century by the Italian anatomists Gabriel Fallopius–who gave his name to the Fallopian tubes–and Realdo Colombo, both of whom claimed to have discovered the organ.[84] In *De Re Anatomica* (1559), Colombo discussed the anatomy of the external female genitalia and a 'certain small part, which is elevated on the apex vaginae above the foramen from which urine exits';

> And this dearest reader is that, it is the principal seat of women's enjoyment in intercourse; so that if you not only rub it with your penis, but even touch it with your little finger, the pleasure causes their seed to flow forth in all directions, swifter than the wind, even if they don't want it to... Since no one else has discerned these processes and their working; if it is permissible to give a name to things discovered by me, it should be called the love or sweetness of Venus. It cannot be said how much I am astonished by so many remarkable anatomists, that they not even have detected [it] on account of so great advantage this so beautiful thing formed by so great art.[85]

Fallopius rejected Colombo's claim to primacy. 'Modern anatomists have entirely neglected it,' he wrote, 'and if others have spoken of it, know that they have taken it from me or my students.'[86] Vesalius was unmoved; he had been Colombo's teacher, but unlike Fallopius and Colombo, he did not see a role for the clitoris, since he viewed the female form as an inverted male.[87] Yet the Renaissance discovery was actually a rediscovery; many years previously Greek, Persian, and Arabic writers had discussed the clitoris, though they disagreed over its function.[88] The names used for the clitoris varied: 'Hippocrates used the term columella or little pillar. Ibn-Sīnā named the clitoris the albatra or virga (rod). The Arab physician and surgeon Abu al-Qasim Khalaf ibn al-Abbas Al-Zahrawi (known in the West as Albucasis) named the clitoris tentigo (meaning tension). 'Amoris dulcedo' (sweetness of love), 'sedes libidinis' (seat of lust), and 'gadfly of Venus' were all terms used by Colombo.'[89]

Increasingly detailed descriptions of the clitoris were put forward in the sixteenth and seventeenth centuries by the Swedish polymath Caspar Bartholin the Elder and the Dutch physician and anatomist Regnier de Graaf. It was De Graaf who insisted on giving the clitoris its formal name to avoid confusion, 'clitoris' being derived from the Greek 'to rub'.[90] De Graaf noted the continued lack of reference to the clitoris in most anatomical work, anatomists making 'no more mention of this part than if it did not exist at all in the universe of nature. [Yet in] . . . every cadaver we have so far dissected we have found it quite perceptible to sight and touch.'[91]

After De Graaf's studies little changed in the discussion of the clitoris until the German anatomist Georg Ludwig Kobelt's account of *The Male and Female Organs of Sexual Arousal in Man and Some Other Mammals* (1844).[92] Kobelt sought to show that male and female organs were 'entirely analogous', though his anatomical knowledge of the female body was 'still full of gaps'.[93] Kobelt performed dissection, comparative anatomy, and injection studies, and his main contribution to anatomical knowledge was an account of the musculature surrounding the clitoris as well as its role in sexual response. One might imagine that this anatomical recognition of the clitoris marked a turning point in understanding the female body. But that has not been the case. In the twentieth century the anatomical literature on the clitoris again was the 'victim of cultural convention', as one historian has put it; 'until relatively recently, detailed diagrams in early editions of well-established anatomy texts were either omitted or replaced with figures with the clitoris unlabeled'.[94]

This trend has been reversed in recent years, most notably with the work of the Australian urologist Helen O'Connell and her Melbourne colleagues who have reappraised the anatomy and function of the clitoris using modern dissection and imaging techniques. O'Connell has shown that medical textbooks remain woefully lacking in detailed information, describing male anatomy fully, yet only noting differences between male and female anatomy rather than providing a full description of female anatomy. O'Connell also observed how inadequate single-plane anatomical illustration is to convey clitoral anatomy. Using Magnetic Resonance Imaging (MRI), which uses magnetic field and radio waves to create detailed images of the body, O'Connell has drawn attention to the depth and complexity of the musculature around the clitoris. Rather than being a delicate knob of tissue below the pubic bone, she has demonstrated, the clitoris consists of a head that is the external component that is attached to a body two to four centimetres long with 'arms' up to nine centimetres long. Those arms give rise to two bulbs on either side of the vagina.

While the penis is external and pendulous, the clitoris should therefore properly be understood as an internal organ, and not part of the external genitalia at all. Instead it is 'as large and significant as the penis'.[95] The clitoris' full structure and capacities are only identifiable through multiple ways of viewing. Scientific lack of recognition for the clitoris as a scientific organ therefore arguably reflects the masculinist bias of science: female sex organs are often regarded only 'as a canal for penetration'.[96] Inadequate sex education, moreover, perpetuates a lack of awareness about the clitoris, along with an additional masculinist bias in teaching that presents the entire male reproductive system on the one hand and the internal female reproductive system (without the clitoris) on the other. O'Connell's findings were first published in 1998, and it has taken several years for them to be transmitted into scientific teaching.[97]

This consistently recurring gap between female anatomy and representation, as well as the ways in which female sexual experience continues to be controlled and monitored and scrutinized throughout the globe, makes the story of the clitoris 'a parable of culture', argues O'Connell; 'of how the body is forged into a shape valuable to civilization despite and not because of itself'.[98] In other words, a lack of interest in, or a fear about female sexuality means that the clitoris has necessarily taken a back seat in anatomical teaching. Like the hymen, which has been politicized for related but separate reasons, the clitoris symbolizes historically persistent fears about female sexuality: disorderly, uncontrollable, and independent, a force for agency above and beyond its medico-scientific and social framework.[99]

As these examples suggest, attempts to anatomize the body in the nineteenth century did not give much agency or control to women's bodies, especially their sexual organs. This is entirely consistent with Victorian ideals about womanhood, femininity, race, and the domestic environment where bourgeois women were supposed to be focusing their energies. Women's roles, especially white, middle-class women's roles, were reproductive and civilizing. We find the same kinds of concern for female sexuality in Shakespeare's time, though the ideal of femininity was less focused on the domestic environment.[100] Nevertheless, in the early modern period, as today, women's bodies were defined by their social and political function. That included the presumption that female emotional characteristics were grounded, like their intellect, in the organs of the body, in the humours and the fluids, and even, as the next chapter will show, in the heart itself.

4

'Soft and Tender' or 'Weighed down by Grief'

The Emotional Heart

The heart (n.)

1. The hollow muscular or otherwise contractile organ, which by its dilatation and contraction, keeps up the circulation of the blood in the vascular system of an animal.
2. Considered as the centre of vital functions: the seat of life; the vital part or principle; hence in some phrases = life.
3. MIND, in the widest sense, including the functions of feeling, volition, and intellect.[1]

THE TRANSPLANTED HEART

In 2006 a British woman, Jennifer Sutton, underwent a heart transplantation operation. She had developed a restrictive cardiomyopathy in her teens, a condition in which the heart is restricted from stretching and filling with blood. Sutton was ill for many years before she received a new heart. Her explanted heart was put on temporary display at the Wellcome Collection in London.[2] In Fig. 5 Sutton is seen confronting her heart for the first time as an outsider, rather than an insider, the explanted object having gone from a live organ beating within her chest to a museum exhibit. Sutton described the experience as 'an emotional

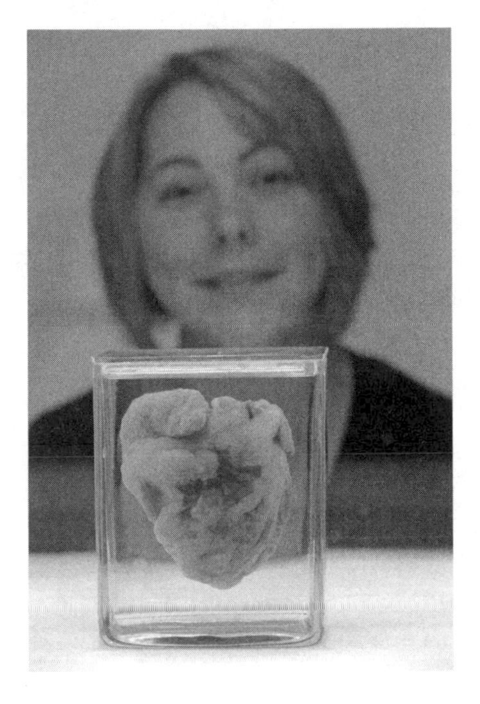

Fig. 5. Jennifer Sutton, photographed with her diseased and explanted heart for an exhibition at the Wellcome Collection in London, 2006.

and surreal' one. 'It caused so much pain and turmoil when it was inside me. Seeing it sitting here is extremely bizarre and very strange . . . Finally I can see this odd looking lump of muscle that has given me so much upset.'[3]

Since the 1950s the transplanted heart has been the subject of much anxiety and upset—as well as amazement and gratitude. This is because of the extraordinary meanings with which the heart as an organ is invested. As this chapter will show, the heart is no mere pump, though it has been regarded that way by surgeons and physicians since the mid nineteenth century, but a space associated with the self, with emotion, with feeling, and with the soul. I have written about the heart's dominance and then decline as an organ of the self between the ancient world and the present day.[4] Here I want to consider that story in relation to the body as a whole, to the process by which the imagery of the heart—not as a lump of muscle on the butcher's slab but in the delicate symmetry we might associate with

Hallmark cards—became an important symbol of emotion, especially love. A focus on the heart, moreover, in medical discourse, as well as in the arts, literature, painting, and music, has consistently affirmed the heart as an organ of the soul.[5] Although it is the brain that has come to dominate discussions of our selves as thinking, feeling beings, the traditional meanings of the heart continue to linger: meanings that derive linguistically and physiologically from ancient theories of the body and the mind.

THE TROUBLING HISTORY OF THE TRANSPLANT

Before the 1960s the notion of a heart transplant arguably inspired as much anxiety as a brain transplant does today, probably because the focus of the 'self' has moved from the heart to the brain as detailed below. But there were other reasons why heart transplants troubled people. The academic and political writer Ali Mazrui wrote about transplants in the wake of Dr Christiaan Barnard's first successful transplant carried out in South Africa.[6] In a climate of racial conflict and inequality, Mazrui's main fear was that black people would be 'spare parts' for whites. But he also drew attention to the meanings of the heart as a moral, emotional, and spiritual symbol. Mazrui asked what it might mean for the inheritor of a heart; might it bring a change of identity and even of soul? 'Would the dead person be envious? . . . Would the beneficiary suffer frightening moments of imaginative confrontation with an accusing finger from the person who left him his heart?'[7]

These kinds of questions remain with us, even though heart transplants are no longer radically new. Today more than 5,000 cardiac transplants take place around the world annually, though nearer to 50,000 people are eligible candidates for transplantation.[8] Many people believe heart transplants bring more to the recipient than just a replacement organ. Claire Sylvia for instance, a dancer who had a heart transplant, claimed that that the new heart changed her personality and emotions and even her tastes, leaving her craving chicken nuggets and beer, favourites of her teenage male donor that Claire had never before enjoyed.[9] There are other similar cases of transplant donor families and recipients being convinced that a person's personality, taste, and temperament are transferred along with the replacement heart.[10]

Why do these ideas coalesce around transplantation? And why does the heart have such lingering emotional meanings? After all, in medical terms the heart is just a pump; a specialized muscle that pumps blood round the body. Each day the average heart beats about 100,000 times. As it works it transports 5,000 gallons of blood round the body, delivering oxygen and nutrients to the cells and carrying away unwanted carbon dioxide and waste products.[11] For some patients a psychoanalytical interpretation has been offered for the feelings associated with transplantation; it is difficult to adjust emotionally to benefiting from a donor organ that is only available because someone died. But not all organs evoke the same anxiety. The transplanted heart invokes feelings different from those generated by any other form of transplant: a lung, say, or a kidney. This is because the heart has special, almost sacred significance in Western culture.[12]

The symbol of the heart is used extensively as a visual symbol, from Valentine's cards to love letters, from graffiti to tattoos. There is a logic to its use: we *feel* emotions in our hearts—our pulses race when we are excited, impassioned, and frightened—and those feelings seem separate from our mental processes. We might feel in our hearts but we know in our heads: a distinction between knowledge and feeling that is part of our Cartesian inheritance.[13] This mind–body split was outlined by the French philosopher René Descartes in a philosophy that has commonly become known as dualism, as opposed to holism, in which mind and body are one. Thus we might privilege our 'hearts' over our 'heads', claiming emotional over rational reasoning as the clue to a better or happier life. Or we might try to reason with our passions, to bring them into line as the stoics recommended.[14]

If we consider the dictionary definitions of 'heart' that introduced this chapter, we find a number of contradictory meanings, the third of which, 'MIND, in the widest sense, including the functions of feeling, volition, and intellect', is the most problematic. This suggests that the heart actually possesses a wisdom and truth that is separate from the brain. This meaning remains part of our linguistic inheritance. The heart speaks to us of things that are *heartfelt* and therefore genuine, possibly heart-warming and compassionate but always true—as in Hamlet's 'heart of heart'.[15] Since before Shakespeare's time the qualities of the heart have been used to indicate a person's temperament and personality: people who are kind are 'warm-hearted', and people who are cruel are 'cold-hearted', 'stony-hearted', or even 'heartless'. This language makes sense within the context of an emotion physiology and a view of the self that is rooted in ancient ideas about the body and the mind.

THE GALENIC INHERITANCE

I have discussed the Galenic model of the body in the introduction. It is important to note than humoral medicine viewed emotions as physical entities, grounded in the material body. While the brain was the seat of reason, and the liver was the site where humours were produced, it was the heart that was the site of emotion or passion. As Robert Burton, the English scholar and incumbent of St Thomas the Martyr in Oxford put it in his most famous work, *The Anatomy of Melancholy* (1621), the heart was 'the seat and foundation of life, of heat, of spirits, of pulse and respiration, the sun of our body, the king and sole commander of it, the seat and organ of all passions and affections'.[16] In this heat-based economy the heart was the point where blood was heated or cooled, respectively, under such passions as anger or fear. This view was remarkably consistent, as seen in a comparison between the description of rage in a conduct book by the French writer Pierre de la Primaudaye, and the preaching of the English cleric John Downame:

> For first of all when the heart is offended, the bloud boyleth round about it, and the heart is puffed up: whereupon followeth a continuall panting and trembling of the heart and breast.[17]
>
> Pierre de la Primaudaye, 1618

> [Anger is] an affection, whereby the bloud about the heart being heated, by the apprehension of some injury offered to a man's self or his friends, and that in turn, or in his opinion onely, the appetite is stirred up to take revenge.[18]
>
> John Downame, 1609

This emphasis on the heart as the agent of heating and concoction was compatible with an emotion physiology that linked body, soul, and mind in a complex union. And because Renaissance physicians regarded the soul and the body as indivisible, the soul was involved in any emotional experience. The English writer Thomas Wright called the passions 'operations of the soule, bordering upon reason and sense, prosecuting some good thing, or flying some ill thing [and] causing there withall some alteration in the body'.[19] The communication of an image via the external senses (especially sight) to the brain preceded any judgement about its value; a subsequent alteration in the body caused 'spirits' to move to the heart, where they would 'signify' the object, and the heart would bend itself to seek or avoid the same.[20] This is what Galen meant when he referred to the heart as being attracted to, or repelled by, a person or thing that was good or bad.

This cardiocentric view of the self and the emotions was not just Western, but shared by Egyptians and Mesopotamians, which is why the weighing of the heart formed such a crucial role in determining one's place in the afterlife according to the *Book of the Dead*.[21] After death the deceased would stand to be judged by Osiris, classically depicted wearing a distinctive crown, with two large ostrich feathers on each side. He also held a symbolic crook and flail as the heart of the deceased was weighed against the feather of truth. Thoth, the god of knowledge, usually pictured with the head of an ibis, recorded the judgement. And the jackal-headed god Anubis tended the scale, ready to devour the deceased if her heart weighed more than the feather (for that indicated she was guilty of sin). If the heart was lighter than the feather then the deceased was allowed to enter the afterlife.[22] So central was the heart to Egyptian beliefs about reason, emotion, memory, and personality that it was usually the only organ left in the body during mummification. All others were removed and either placed in canopic jars for the afterlife, or discarded like the brain, removed through the nose with a hook.[23] The British surgeon and antiquarian Thomas Pettigrew (1791–1865), sometimes known as 'Mummy' Pettigrew—was famous for unwrapping and anatomizing mummies during private parties at his London residence. In his writings he explained how the brain was 'extracted through the nostrils, which was effected by the aid of iron (bronze) crochets'.[24]

The Greeks had a more divided view on the human body. Some held a cardiocentric perspective and others were cephalocentric, meaning that they focused on the brain.[25] Much of Galen's anatomical work, notably on pigs, was concerned with the structure and function of the brain. Nevertheless, the heart was still crucial because of the ways it influenced passions and the soul. The soul remained an important part of early modern physiology, just as it had been for the classical and the mediaeval worlds. However, up to three souls were thought to exist and the soul had the power to 'excite Corporeal Passions directly', in the words of the English writer Walter Charleton.[26] Today it is perhaps hard to imagine a world in which the soul was so heavily involved in physiological processes rather than an abstract spiritual idea. The early modern soul summoned humours like melancholy for pain and sadness; blood and choler for anger. The agitation of the spirits in the emotion of joy and the free flow of the blood throughout the body were in direct contrast to the physiological experience of fear, when the blood retreated and the soul shrank away from any threat. This process explained the physical manifestations of emotion, the hair sticking up on end, the flushed face, even the gnashing teeth of rage. As Downame's *Treatise of Anger* explained of anger, the passion:

> Maketh the haire to stand on end, shewing the obdurate inflexiblenesse of the minde. The eyes to stare and candle, as though with the Cockatrice they would kill with their lookes. The teeth to gnash like a furious Bore. The face now red, and soon after pale, as if either it blushed for shame of the mind's follie, or envied others good. The tongue to stammer, as being not ablc to expresse the rage of the hart. The bloud ready to burst out of the vaines, as though it were affraide to stay in so furious a body. The brest to swell, as being not large enough to containe their anger, and therefore seeketh to ease it selfe, by sending out hot-breathing sighes. The hands to beate the tables and walles, which never offended them. The joyntes to tremble and shake, as if they were afraid of the mindes furie. The feete to stamp the guiltlesse earth, as though there were not room enough for it in the whole element of the aire, and therefore sought entrance into the earth also. So that anger deformeth the body from the hayre of the head to the soale of the foote.[27]

If anger caused the blood to boil around the heart, the reverse physiological process was associated with fear, as with grief and sorrow. These 'negative' emotions caused the soul to contract so that the animal spirits were:

> recalled inward, but slowly and without violence: so that the blood being by degrees destitute of a sufficient influx of them, is transmitted with too slow a motion. Whence the pulse is rendered little, slow, rare and weak, and there is felt about the heart a certain oppressive strictness as if the orifices of it were drawn together, with a manifest chilness congealing the blood and communicating itself to the rest of the body.[28]

Such 'dejecting symptoms' had a long-term detrimental effect on the health, since sadness as well as anger:

> obscures the judgement, blunts the memory, and in a word beclouds the Lucid part of the Soul: it doth moreover incrassate the blood by refrigeration, and by that reason immoderately constringe the heart, cause the lamp of life to burn weakly and dimly, corrupt the nutritive juice and convert it into that Devil of a Humour, Melancholy.[29]

This humoral approach provided an explanation for the physical symptoms of emotion that would subsequently be perceived by Charles Darwin as evidence of evolutionary development.[30] The experience of the mind and body, emotional differences between people (which translated into national and ethnic as well as age and gender differences), and health and disease were neatly accommodated into a holistic account of what it was to be human.

This vision began to slip by the late eighteenth century, though the medical therapeutics that had evolved from it, including cupping and bleeding and purging to remove bad humours, remained in place well into the nineteenth century. There were two main changes that made a difference to that narrative: firstly, the discovery of blood circulation, which impacted on the belief that humours could congregate in different parts of the body, and secondly, seventeenth-century philosophical ideas that separated the body from the mind.

For Galen the function of the arteries, like that of the lungs, was to cool the heart and regulate the temperature of the body. The heart was a heating agent as well as a spiritual organ; its ability to contract and enlarge was evidence of the soul at work in attracting good and repelling evil.[31] There were challenges to this model, especially from the Arab world, well before the seventeenth century, though they have been largely unacknowledged by the Western medical tradition until recently. The thirteenth-century Arabic physician Ibn-al-Nafis and the sixteenth-century Spanish theologian Michael Servetus both considered the possibility of pulmonary circulation. The work of Ibn-al-Nafis deserves to be more widely acknowledged as a great achievement and part of the large contribution of the Arab world to modern civilization, science, and medicine.[32] There were others in the sixteenth century who worked on the circulation of the blood, including the Italian anatomist Realdo Colombo, the Italian physician Andrea Cesalpino and the influential Belgian anatomist Andreas Vesalius.[33] Yet it is William Harvey who has gone down in history as the first proponent of blood circulation and the 'father' of the vascular and circulatory systems.[34] Harvey published his *Exercitatio anatomica de motu cordis et sanguinis in animalibus* (*On the Motion of the Heart and Blood*) in 1628 in Frankfurt. This book, the first Western account of the action of the heart and the movement of the blood round the body, was dedicated to King Charles I.[35]

Harvey was born in Kent in 1578 and studied at Cambridge and then at the University of Padua in Italy, the centre of western European medical instruction. Back in London, Harvey joined the College of Physicians and married the daughter of Lancelot Browne, physician to King James I. Harvey subsequently became court physician to James I and then to Charles I. A physician at St. Bartholomew's Hospital and Lumleian Lecturer at the College of Physicians, Harvey used experimental dissections and vivisection to inform his discussion of blood circulation. He rejected Galen's idea that the liver made the blood, which was used up by the body. Through experimentation Harvey argued that the liver would need to make 540 pounds of blood every hour for this to be correct. Instead, he argued, blood was recycled by the body, and that it flowed through the body in two loops; one which went to the lungs and received oxygen and another that distributed the oxygen to the organs and body tissues.[36]

Harvey's work needs to be situated in the context of Renaissance experimentation, when anatomical dissection was demystifying the body and making its processes more explicable to the gaze of science.

Moreover, when accompanied by the ideas of philosophers like Descartes, it was possible to imagine the heart as governed not by the soul, but by secular material processes.[37] More specifically, Descartes developed the 'reflex' to explain how body and brain acted together.[38] That does not mean he had no role for the soul: rather, he gave the soul a precise location in the brain, in the pineal gland, where it moved between mind and body. Incidentally this was in contrast to Harvey, who argued that the soul was a property of the blood, an influence taken directly from Aristotle. Moving the soul to the brain rather than the heart was important. It sparked the gradual secularization of the heart, the process by which it became possible to view the heart not as a sacred object but a mere pump, subject to disease and decay like any other physical organ.

THE HEART AS MATERIAL ORGAN: FROM FURNACE TO PUMP

The discovery of blood circulation effectively critiqued the physical basis of humoralism. Galen believed that blood moved around the body by passing between the ventricles through means of invisible pores where the venous and arterial systems came into contact. It would be many years before that theory was disputed.[39] One of the most important aspects of humoral medicine to note, given Galen's limited research into human cadavers, is that physiological theory drew from a broad metaphysical framework. Less emphasis was placed on objective experimentation, which became a touchstone of scientific investigation in the long nineteenth century.

Even though the work of Harvey and Descartes heralded a new age of the heart and the brain in seventeenth-century medicine and science, a period of pathological anatomy and dissection, of knowing and measuring and classifying the human body according to distinct systems and principles, humoral medicine continued at a practical level well into the nineteenth century. This was most evident in treatments like bloodletting, and a continued belief in the non-naturals as well as in pre-pathogen attitudes towards illness as imbalance. It has become unfashionable to speak of a seventeenth-century 'revolution' having taken place in science.[40] Historians have asked whether the institutions and practices of the period—including natural philosophy, natural history, astronomy, mechanics, anatomy, medicine, astrology, and alchemy—were sufficiently similar to modern science to warrant the label 'scientific'. 'Natural philosophy' is probably the closest thing to what we might consider physical science.[41] Regardless of semantics, medical conceptions of the heart as a pump became commonplace after Harvey, when machines of all kinds were invented.[42] And yet Harvey spent more time comparing the circulation of the blood to the weather cycle than to its function as a pump. There is strong evidence in Harvey's work of the influence of Aristotle and natural magic, principally astrology and alchemy.[43]

The metaphors used in science and medicine matter. The language in which the mind–body relationship has historically been conceptualized is indicative of broader cultural shifts in social and economic life. Thus metaphors of clockwork bodies (and hearts) became conceivable at the same time as manufacturing mechanisms made such phenomena part of the material world. A similar process can be seen in the early twenty-first century shift towards seeing the mind as a complex filing system, processing information like a computer. Memories and experiences exist in separate files to be accessed when required by the mainframe operator. Earlier influential metaphors include the mind as a 'filing cabinet' and the body as a hydraulic kettle that boils until emotions are released.[44] In cultural terms the heart still aches and breaks over lost love; we say that people have hearts of stone or hearts of gold; our emotions 'weigh us down', unless we have the heart of a lion; our hearts are often 'in our mouths' as we anxiously await news; they 'sink' when we are sad but 'soar' when we are happy. Some of us wear our hearts on our sleeves, even when we know 'in our heart of hearts' that we should not. We set our hearts on things and people, we speak from the heart, we have big hearts, hard hearts, and cold hearts. In symbolic and linguistic terms we are a long way from viewing the heart merely as a pump. Moreover, at the same time that the heart was viewed as an entirely secular organ in the peak of the Enlightenment period, there was something of a backlash. It is fair to say that, as the French thinker Michel Foucault has suggested in his work on counter-memories, all dominant narratives have counter-narratives.[45] The counter-narrative to the heart of science was the heart of the Romantics.

THE HEART OF ROMANTICISM

Amidst the empiricism and rationalism of the Enlightenment the heart was revitalized by the Romantic poets as a source of the truth and the divine, of the natural, as opposed to the new industrial age. From the late eighteenth to the mid nineteenth century, the Romantic influence was felt across all artistic disciplines and several continents. Reactions against neoclassicism, order, restraint, and, above all else, reason was manifested in a love of individualism and the passions, and in an interest in the mystical and the supernatural. In Britain, William Wordsworth, Samuel Taylor Coleridge, Percy Bysshe Shelley, George Gordon, Lord Byron, and John Keats propelled the English Romantic movement, which crossed the Atlantic through the work of American poets like Walt Whitman and Edgar Allan Poe.[46]

The importance of the heart of feeling to Romanticism was represented rather touchingly in the probably apocryphal tale of the poet Shelley.[47] After Shelley's death by drowning, at the age of only twenty-nine, his body was cremated in the presence of his friends and fellow writers Edward John Trelawny and Leigh Hunt. It was reported that Shelley's heart would not burn, that it was retrieved from the flames by his friend Trelawny, and passed—presumably along with his love, his passion, and his poetic spirit—to his wife Mary Wollstonecraft Shelley, the author of *Frankenstein*, who kept it wrapped in silk until her own death.

If the history of the Romantic heart is well known, less so is the association between creative souls and heart disease. We are used to the resonance of tuberculosis during the Romantic period, but heart disease also emerged as a significant problem, especially among middle-class, literary individuals.[48] This phenomenon is seen clearly in the life and work of Harriet Martineau, the British writer and philosopher who struggled for most of her life with health problems.[49] During Martineau's life there was a renewed shift towards the heart in literary culture, both in terms of heart-centred imagery, and in links with emotion and authenticity.[50] In fictional writings, including those by Martineau herself, the heart functioned as a symbol of intense sensibility and feelings. Along with the writings of Elizabeth Barrett Browning (Sonnet V of *Sonnets from the Portuguese* begins: 'I lift my heart up solemnly') and of Christina Rossetti ('My heart is like a singing bird'), Martineau's novel *Deerbrook* (1839) is filled with references to the emotional heart.[51]

Deerbrook's subject concerns the fortunes of the Ibbotson sisters, Hester and Margaret, who arrive at the village of Deerbrook to stay with their cousin, Mr Grey, and his wife. Margaret attracts the attention of the local medical practitioner, Edward Hope, though he is persuaded to marry the beautiful Hester. It is a miserable marriage, overshadowed by Hester's jealousy and a series of misunderstandings between Margaret and her own lover. Hope's fortunes take a turn for the worse when he is accused of grave-robbing, and he succumbs to a near fatal fever. Eventually, and according to literary convention, health and order are rehabilitated, along with the marriage of Hester and Edward.

Described by many critics as one of the first 'domestic novels' of the Victorian era, *Deerbrook* provides a sentimentalized account of the meanings of love and affection between siblings, acquaintances, and lovers. Central to the novel's symbolism of love, authenticity, and choice, hearts possess a vitality and morality of their own. They 'dance' and 'sink' under extreme emotions. Individuals are 'heavy' at heart when grief-stricken, but have 'cheerful' hearts when optimistic. Some women possess 'kind' hearts, and concern for others, while others' hearts are 'hard', and immune to their plight. Over time, hearts became softer or harder, depending on experience. As expressed in the novel's denouement, the male protagonist 'had really gone through a great deal of anxiety and suffering lately, and his heart was very soft and tender just now'.[52]

Yet hearts were also connected to the mind in *Deerbrook*, and to reason. The possession of a 'heart and a conscience' was important to humanity, and 'sympathy within [one's] heart and mind', ideal. Hearts also possessed knowledge that was unmediated by human error. To know oneself, or one's subject was to 'learn by heart'; to follow one's truth was to 'follow' one's heart. As emotional receptacles, hearts were filled or emptied by degrees of feeling, and characters became 'heart full' or 'single-hearted' according to the object of their affections. Hearts also changed shape, depending on the emotion that they expressed. As in seventeenth-century discourses on emotion physiology, hearts 'swelled' when they had secrets to impart; and sometimes 'inflated' with pride.[53]

The truthfulness of the heart was physically retained in its structure, for it embodied memories and experiences; it remembered words and tones of speech. Questions 'struggled' in the heart, and the heart physically resisted the suppression or avoidance of its truth. To live untruthfully, and to be distressed 'at heart', was to invite unhappiness or even disease. Little wonder, then, that the actions and sensations of the heart were subject to a series of specific physiological effects. Hearts became 'heavy' and 'dismayed', 'sick' or 'affected' by experience, so that they 'leapt up' (in joy) or 'trembled' and 'beat' in fear and anticipation. When extremely distressed, hearts 'throbbed' painfully, or were 'weighed down by grief'.[54] Any extreme weight afflicting the heart, such as depression and disappointment, constricted one's breathing and caused breathlessness and palpitations, symptoms associated with heart disease. Beyond its physical structure, hearts signified intimacy; to metaphorically share the contents of one's heart was the opposite of loneliness, when the heart felt 'wringed' in grief. Hearts were sociable organs; they needed nurture to stay well. Without kindness and affection (indeed, without intimacy), hearts weakened and became sick. How ironic, then, as one character exclaimed in *Deerbrook*, that 'we cannot see into one another's hearts'. For on some level we must always be alone; 'what lies deepest in [our] heart' is ultimately impenetrable by others.[55]

The weakening and disease of the heart was a much debated topic among the medical profession and the educated classes of Victorian Britain. Popular medical encyclopaedias, journals, and newspapers dealt with the diagnosis of heart disease, which coincided with the growth in cardiac medicine as a clinical specialism.[56] There was a rapid growth in specialist physicians and hospitals, and an apparent rise in heart disease as a cause of death.[57] This concern in a rising death rate as a result of cardiac dysfunction was both widespread and believed to reflect the rapid pace of modern life in Victorian Britain. Certain types of heart disease, such as angina pectoris, were found more often in men. Yet the literary imagery of heart disease as an enfeebling and emotional disease was peculiarly female. In the mid nineteenth century, heart disease acquired a certain status among female literati who combined literary acclaim with fragile health and poetic artistry. This is seen in accounts of writers who fretted over their own hearts in letters and diaries. They perhaps positioned themselves—wittingly or unwittingly—as cardiac sufferers because of the association of heart disease with sensitivity and depth of feeling.[58]

Today, heart conditions are more likely to be associated with high cholesterol, or stress and anxiety, than with acute sensitivity. They are also more likely to be associated with obese and poorer people, and with men, than in Martineau's time. And yet the association of the emotions and the heart remains intact. Moreover, in terms of the emotional impact of heart disease, and the impact of emotions on heart health, a symbiotic relationship is increasingly accepted. There is, for instance, evidence that anxiety and stress are seen as causal factors in the development of chronic heart conditions.[59] Martin Cowie, Professor of Cardiology at the Brompton Hospital, has found there is an increased risk of dying in the six months after bereavement, especially among widowers. This is partly because of accidents caused by distraction, but it is also explained by the incidence of heart attacks and strokes resulting from hormonal surges accompanying grief and stress.[60]

TOWARDS A NEW UNDERSTANDING OF THE EMOTIONAL HEART OF SCIENCE

Until the late twentieth century references to the emotional heart, especially in transplantation cases, were rhetorical and anecdotal. The dominant medical framework simply did not allow the heart to be more than a functional and material object. Common sense suggested that the heart was complicit in our emotional life; after all, we do feel emotions in the heart. And many of us on a 'gut' level (an important reference to intuitive knowledge that I will return to elsewhere) might have reservations about acquiring the heart of another. Might it change us as people? Would we feel, think, or remember differently? Since the emergence of pathological anatomy and cardiology there has been a disjuncture between the scientific and emotional hearts. Yet there is evidence that a new model is emerging, one that combines both science and the emotions in its very structure.

One traditional response by surgeons to discussions of the spiritual or emotional heart has been denial: the heart has no special meaning. This attitude is understandable: how else could one operate without a sense of dread? That was one surgeon's response when invited by me to attend a workshop on the spiritual significance of the heart.[61] Medical students, too, have expressed feelings of trepidation when holding a heart, as though it might contain the essence of a person's soul.[62] This is an important phenomenon to recognize; in most medical schools students have the opportunity to dissect both the heart and the brain, but there has been little research into the metaphysical views of those students and how they might feel about dissections.

A recent investigation into body donation did ask these questions and followed students into the dissection room to explore their attitudes. Of the sixteen students interviewed (8 male and 8 female), ten described themselves as religious, either Christian or Jewish. Some of the students were unperturbed while dissecting the brain, because they held that the heart was the 'seat of the soul'.[63] Students referred to the heart as 'amazing' and 'interesting', even treating its dissection as a privilege:

> Well removing the heart... and holding the heart was like holding the seat of the soul... there was some sort of aura involved in it and it was like um just holding the human heart, removing it from the cadaver like that was... um, you know it was different. It had... it felt like... 'cos to me, I feel like the heart is one of the most important, well I think it is, it seems like the most emotionally involved part of the body, like everyone always talks about 'if your heart's in it', you know it's like... it's the most... it's the strongest organ of the body I think. Um so removing the heart had some kind of aura involved around it.[64]

The reactions of medical students reminds us that anatomy is not merely a biomedical discipline, but also a philosophical endeavour.

Since the 1990s, new scientific languages have developed to address the continued existence of the sentimental heart. Central to those languages is the belief in 'cellular memory', and the proposal that the heart possesses a 'little brain' that remembers and feels in its own right. Research into the heart's brain is associated most explicitly with the American physician Andrew Armour and the Institute of Heartmath, and is not yet considered mainstream.[65] Armour and his team claim that the heart like the brain contains an intricate network of neurons, transmitters, proteins, and support cells. Its elaborate circuitry, which can detect hormones and neurochemicals as well as pressure and impulses associated with emotions, might enable the heart to act independently of the cranial brain: to learn, remember, and even feel and sense.[66] What is interesting to me is not whether or not the concept of cellular memory is any more or less objectively 'true' than other medical explanations, but the ways in which scientific frameworks evolve and are accepted or rejected according to prevailing metaphysical and scientific frameworks. The concept of the heart's 'little brain', which is active and not just mechanistic, allows for a soul-like presence to be associated with the heart. It might also accommodate anecdotal transplant memories into an avowedly scientific discourse.

Who knows where these heart and mind reformulations will take us, though the heart as an emotional organ looks set to prevail—at least at the level of popular culture. The heart taken from Jennifer Sutton's chest looks nothing like the Hallmark heart we find on Valentine cards. There are numerous suggestions as to the origins of the romanticized ideal: the redness symbolizes blood and passion, for example, while the shape resembles the courting ritual of swans' necks, or perhaps the silphium seed, used as a contraceptive since classical antiquity. It has even been claimed that the heart is a signifier for a woman's buttocks or her parted vulva.[67] Whatever its origin, the symbol is far more aesthetically pleasing than the alternatives, even if we now believe that emotions reside in the brain. After all, 'I brain you' is far from romantic.

5

Mind the Brain

From 'Cold Wet Matter' to the Motherboard

The brain is the source of all the feelings, ideas, affections and passions; their manifestations, therefore, must depend on the brain and be modified by it.'

(Francis Gall, *On the Functions of the Brain*, 1835[1])

On September 13, 1848, Phineas P. Gage, a 25-year-old construction foreman from New Hampshire, was directing a work gang blasting rock for a railway south of Vermont.[2] In order to lay the track, the terrain had to be levelled by controlled blasting. One of Gage's tasks was to perform the detonations. He drilled holes into the rock before they were filled in with explosive powder and sand, and then he used a fuse and a tamping iron to trigger an explosion. Gage had performed these actions numerous times before. On this occasion he asked his assistant to place the sand and then turned his head, perhaps momentarily distracted by another colleague. When he turned back he tamped on the powder, not realizing that the assistant had not yet covered it with sand. The powder was ignited and a powerful blast sent Gage's tamping iron, a pointed metal rod measuring three feet seven inches in length, one and one-quarter inch in diameter (more than a metre long and three centimetres in diameter), towards Gage's face. The power of the explosion sent the rod through Gage's cheek, his skull and his brain. It exited through the top of Gage's head, passing straight behind the back of one eye before landing on the ground eighty feet (about twenty-five metres) away, smeared with blood and brain. His shocked workmates observed that Gage momentarily lost consciousness, falling on his back with a 'few convulsive motions'.[3] Within a few minutes, however, Gage was back on his feet, slightly dazed but talking and walking with little assistance. He was carried upright in an oxcart for three-quarters of a mile (about one kilometre) back to his lodgings.

Gage might have died from his injuries and disappeared from history. Instead he survived. But he was changed forever; not in terms of what he could physically do, as his motor functions and his physical strength seemed unchanged. Nor had his intelligence been lowered; he had no impairment in hearing or speech and he could learn new material as easily as before. But his personality, emotions, and social behaviour had radically altered. Gage's local doctor, John Martyn Harlow first discussed Gage's case in a paper published in the *Boston Medical and Surgical Journal* in 1848, with a short follow-up note the following year.[4] Twenty years later he published a final paper recounting Gage's subsequent history. Gage's case provides a unique vantage point from which we might explore the ways the brain has been viewed as an organ of both the psyche and the soma.

Today the brain is understood not only as the centre of the nervous system, but also as the repository of 'mind', that essence of humanity that is sometimes still regarded separately from the brain and which gives rise to individual thoughts, feelings, memories and beliefs. Despite the cultural resonance of the heart, we live in a neurocentric age. It is the brain that has become the 'dominant framework' through which we understand mind, self, and society.[5] The historian of science Fernando Vidal has termed this process the birth of the 'cerebral subject'; one that proposes identity and the brain are the same thing, and that the brain is the only organ that matters 'in order to be ourselves'.[6] And yet the relationship of our brains to our selves is a complex one. If you ask the average person to point to their minds, they will point to their heads; if you then ask them to point to their soul, they will usually point to their chests. This distinction between soul and mind has become as commonplace as the association between mind and brain, although the latter were once separate entities.

For most of history, 'mind' has been something above and beyond the brain: an independent entity often conceived as the soul, which acted in the spirits, the blood, or the ventricles of the brain.[7] As with other organs, perceptions of the human brain were linked to broader socio-economic and political debates about how the physical body and the body politic functioned. From the time 'reason' emerged as a principle of mind that differentiated humans from animals (*c.*1100 to *c.*750 BCE in Greece), it has been juxtaposed with passion; the head against the heart model discussed in the previous chapter.[8] Reason was proof of intellect and as Aristotle put it, 'intellect more than anything else *is* man'.[9] The unravelling of reason as a result of jealousy in Shakespeare's *Othello* is an instance of this: 'Are his wits safe? Is he not light of brain?' Ludovico asks Iago, shortly before Othello's world falls apart.[10] While the heart is dominant in the language of feeling, therefore, it is the brain and its functions that bear the stamp of the creator, and are used to show the differences between sanity and insanity, human and animals. After all, it was the head that was closest to the heavens in the great chain of being and the brain that was the crown of the redefined nervous system of the Renaissance anatomist Andreas Vesalius. Little wonder, perhaps, that the separation of head and body was central to the symbolic and literal decapitation of power in the English and French revolutions.[11]

It was not always the brain that was the material repository of mind, as discussed below. Yet Gage's accident became a challenging case study in brain specialization and localization; an example of the ways mind was linked to the material structures of the brain. His physician, Harlow, whose tale was initially dismissed as a 'Yankee fiction', was first interested in the case not in relation to such broad questions, but as 'a beautiful display of the recuperative powers of nature'.[12] Certainly he was interested enough to obtain Gage's skull when he subsequently heard of his patient's death.

Harlow examined both the skull and the rod in detail before he deposited them in the Museum of the Medical Department of Harvard University in Boston.

THE PARTIAL RECOVERY OF PHINEAS GAGE

Dr Harlow was at Gage's lodgings an hour after the accident. He helped Gage up the long staircase to his room in the lodgings that were owned by Joseph Adams, the local Justice of the Peace. There, Harlow put his patient to bed. Gage was 'perfectly conscious' and lucid enough to describe the accident, but he was exhausted, the loss of blood from his head such that 'his person and the bed in which he lay was one gore of blood'.[13] The first thing that Harlow and his fellow physician, Dr Edwards Williams, did was to assess and tend to Gage's head. The subsequent description of Gage's injury is striking, especially in relation to the severity of the wound and how much of the brain had been exposed:

> From the appearance of the wound in the top of the head, the fragments of bone being lifted up, the brain protruding from the opening and hanging in shreds upon the hair it was evident that the opening in the skull was occasioned by some force acting from below upward, having very much the shape of an inverted funnel, the edges of the scalp everted [turned inside out] and the frontal bone extensively fractured, leaving an irregular oblong opening in the skull of two by three and one-half inches. The globe of the left eye was protruded from its orbit by one-half its diameter, and the left side of the face was more prominent than the right side. The pulsations of the brain were distinctly seen and felt.[14]

Harlow investigated the wound more fully, sweeping his finger through the hole in Gage's skull to search for bits of foreign matter, though Gage was scarce able to feel it. This was before the days of germ theory; neither of the doctors knew that bacterial infection of the brain was a potential hazard. The doctors set about tidying up the wound, shaving Gage's head before removing the fragments of bone as well as an ounce of brain that protruded through Gage's skull. The larger pieces of skull were replaced 'as approximately as possible' and the wound was dressed with a compress and a roller to keep the dressing tight, and covered with a nightcap. Gage's face, hands and arms were badly burned from the explosion, so they were also dressed before he was left to rest in an upright position.

Gage's mother and uncle visited, presumably to care for Gage, after he spent an uncomfortable night, vomiting and bleeding profusely. He remained lucid, however, and fretted about who was replacing him as foreman, as well as wishing to return to work as soon as possible. Two days after the accident his condition worsened and he 'lost control of his mind and became decidedly delirious'. The doctor passed a 'metallic probe' into the opening of the head, 'down until it reached the base of the skull, without resistance or pain, the brain not being sensitive'. The following day a foul-smelling discharge with blood and pus began to ooze from the wound, 'with particles of brain intermingled, finding its way out from the opening in the top of head and also from the one in the base of the skull into the mouth'.[15]

Poor Gage's condition worsened considerably before he recovered. Ten days after the accident he lost vision in the affected eye. He had a high temperature and was restless and incoherent, convinced that he was dying. Gage's head wound was cleaned and dressed three times a day, always with water and disinfectant and the top of his head carefully covered with oiled silk beneath the wet compresses. Over the next three days his left eye began to stick out even further, 'with fungus pushing out from the internal canthus'.[16] Fungus also pushed its way out of the wound at the top of Gage's head. He could speak in monosyllables and took very little nourishment. Gage was in such a poor state that his doctor and friends and family expected him to die. They prepared his coffin and funeral clothes and Harlow had to fight to keep treating his patient:

> One of the attendants implored me not to do anything more for him, as it would only prolong his sufferings—that if I would only keep away and let

> him alone, he would die. She said he appeared [to have] like 'water on the brain'. I said it is not water, but matter that is killing the man—so with a pair of curved scissors I cut off the fungi which were sprouting out from the top of the brain and filling the opening, and made free application of caustic to them. With a scalpel I laid open the integuments, between the opening and the roots of the nose, and immediately there were discharged eight ounces of ill-conditioned pus, with blood, and excessively foetid.[17]

With the fungus removed the discharge continued to be copious and foul-smelling for a further eight days, from which time Gage's health seemed to improve. On October 6, twenty-three days after the injury, he called for his trousers and asked to be helped out of bed so that he could return to work, though he couldn't raise his head from his pillow unassisted. Just over a week later the fungus was abating and there was 'laudable' or good pus coming from the wound. Gradually Gage's 'sensorial powers' were improving and his mind was 'somewhat clearer, but very childish'.

Between October 1848 and January 1849 Gage made a halting but definite recovery. The doctor's notes show that his treatment was largely traditional. To restore health it was necessary to realign the humours, to get rid of the unhealthy built-up matter in the wound and the body. Gage was believed to have made his first significant recovery once pus had been removed and the second when he was bled from his arm and purged as soon as he was well enough. He then spent the winter 'improving in flesh and strength' and the following April visited Harlow. The doctor found that he had a good general appearance; that he stood quite erect and he walked well. His vision never recovered in the damaged eye and the left side of his face was partially paralysed, but the doctor was 'inclined to say that he has recovered'. Other than a 'queer feeling' in his head, Gage felt well and ready to return to work. He reapplied for his former position as a foreman and it seems that he went back to work for a time but his contractors, who had always regarded him as 'the most efficient and capable foreman', refused to keep him on. The reason they gave was that his mind had been 'radically changed':

> The equilibrium or balance, so to speak, between his intellectual faculties and animal propensities, seems to have been destroyed. He is fitful, irreverent, indulging at times in the grossest profanity (which was not previously his custom), manifesting but little deference for his fellows, impatient of restraint or advice when it conflicts with his desires, at times pertinaciously obstinate, yet capricious and vacillating, devising many plans of future operation, which are no sooner arranged than they are abandoned in turn for others appearing more feasible.[18]

Or, as Harlow put it: 'Gage was "no longer Gage".' He never regained his previous independence and status, apparently preferring to spend his time with dogs and horses than with people. He left New England in 1852, taking with him his tamping iron, for which he expressed some 'affection', since it had been made to his specifications by a local blacksmith. Gage worked in Chile for eight years as a coachman and horseman. He seemed to stay in good physical health until 1859, when he declined. In retrospect, Harlow suggested, this was a result of the deterioration of his brain. In 1861 Gage started experiencing epileptic fits. He was bled by physicians on 18 May but, after a particularly bad series of convulsions, he died three days later.

Twenty years after Gage's death, armed only with the dead man's skull and his tamping iron, Harlow set about trying to understand what had happened to Gage's brain, and why his personality underwent such change. Harlow's interest came at a time when similar cases of neurological damage were being well publicized. In the 1860s the idea of localization came to the fore as different areas of the brain were identified with specific processes. The French surgeon and anatomist Paul Broca and the German physician Carl Wernicke both produced insights into the functional specialization of the human brain in relation to motor function, sensory perception, and language.[19] Gage's case suggested there might be something else at work besides these capabilities; could there be structures in the brain that were dedicated to personality, social functions, and reasonable behaviour?

We can only imagine Harlow's frustration that he did not have access to Gage's brain. He 'regretted that an autopsy could not have been had, so that the precise condition' of that organ could be established.[20] Despite his close following of Gage's travels in North and South America, Harlow did not discover his patient had died until five years after his funeral, at which point he wrote to Gage's parents and requested the exhumation of his skull. By that stage it was not Harlow but the more established surgeon Dr Henry Jacob Bigelow, Professor of Surgery at Harvard University, who brought attention to Gage's case. Harlow told Bigelow about Gage soon after the accident, perhaps because he wanted Bigelow to confirm that these unlikely events had taken place. Bigelow received signed statements from many of the people involved, including the Justice of the Peace in whose home Gage had been lodging. Bigelow wanted to examine Gage himself as soon as he was well enough and invited him to visit Boston for a few days, where Bigelow arranged portraits, made a plaster cast of Gage's head and presented him at the Boston Society for Medical Improvement before writing his own paper that he published in the *American Journal of Medical Sciences* (see Fig. 6).[21]

Bigelow acknowledged that the case was extraordinary; the idea that a man could have a metal rod pass through his head and still walk and talk and know his own name sounded like 'the sort of accident that happens in the Pantomime at the theatre but not elsewhere'.[22] Yet Bigelow was convinced. What was remarkable for Bigelow was the 'singular chance' of the specific trajectory of the rod, which saved the brain from even worse damage. This was not the first time, Bigelow observed, that a part of the brain had been removed without impairing its overall functions; 'atrophy of an entire cerebral hemisphere has also been recorded'. Despite Bigelow's confirmation that the accident had indeed happened, and that Gage had recovered, he did not support Harlow's suggestion that Gage was mentally changed by it. Harlow might have pushed his claims further, but there was no autopsy evidence, which would explain why Gage's case

Fig. 6. Phineas Gage after his accident, posing with his tamping iron.

did not make the same impact as work by Broca and Wernicke. Moreover, the idea that a particular part of the brain was responsible for emotions and the regulation of social behaviour was new, whereas there was already a neural basis for movement and language. Without material evidence other than the skull and a plaster cast, Harlow's findings received no further backing.

That all changed when the Scottish physiologist Sir David Ferrier gave his 'Goulstonian Lectures' in 1878. On the basis of experimental physiology, Ferrier argued, it was clear that there were particular regions of the brain responsible for definite mental functions; brain damage would produce different characteristics depending on the nature of the injury. His removal of pre-frontal lobes in monkeys revealed no physiological changes but distinct character and behavioural alteration, including a loss of 'attentive and intelligent observation'. Harlow's claims were finally upheld; the tamping rod had damaged Gage's pre-frontal cortex but spared the language and motor regions, which explained why he appeared to make a full recovery but suffered some kind of 'mental degradation'.[23]

These observed changes in Gage's personality were consistent with modern analyses of damage to the orbitofrontal cortex.[24] Today, though Gage's accident still encourages debate, it is well established that the frontal cortex is a site for the organization of behaviour and short-term memory, motor attention, and inhibitory control.[25] We also presume that 'mind' largely equals brain. This was not always the case. Since classical times there have been debates over how far 'mind' was separate from the brain, and whether there was an extra and immaterial substance at work; those arguments are not settled today.[26] Moreover, where the soul might have been situated—in the blood, the spirits, the heart, or even the pineal gland—was a question framed in many different ways before the nineteenth century, as seen in the history of the brain as an anatomical object.

DISSECTING THE ONION

As suggested in the previous chapter, the brain has vied for supremacy with the heart for centuries. Before 500 BCE Ancient Greeks were divided between cardiocentric and craniocentric models of the body. While Aristotle viewed the brain as subordinate to the heart, many others—including his teacher Plato—did not.[27] Alkmaion of Croton, one of the most eminent natural philosophers and medical theorists, saw the mind itself as the product of material processes. This is sufficiently close to modern interpretations for neuroscientists to view Alkmaion's claim as an unacknowledged 'revolution in human knowledge comparable to that of Copernicus and of Darwin'.[28] Plato accepted Alkmaion's argument for the immortality of the soul, viewing the brain as the seat of intelligence. Hippocrates, too, stressed the brain's relevance to health and disease as well as personality.[29]

Not only did Hippocrates believe that disorders like epilepsy originated in material brain defects, but also that without our brains we would not be able to perceive, feel, and experience the world, make judgements, or experience pleasure and pain. He went even further, suggesting that it was the health of the brain that determined the health of our bodies and our minds:

> Men ought to know that from nothing else but the brain come joys, delights, laughter and sports, and sorrows, griefs, despondency, and lamentations. And by this, in an especial manner, we acquire wisdom and knowledge, and see and hear, and know what are foul and what are fair, what are bad and what are good, what are sweet, and what unsavory; some we discriminate by habit, and some we perceive by their utility. By this we distinguish objects of relish and disrelish, according to the seasons; and the same things do not always please us. And by the same organ we become mad and delirious, and fears and terrors assail us . . . all these things we endure from the brain. In these ways I am of the opinion that the brain exercises the greatest power in man.[30]

There is no evidence that Hippocrates conducted autopsies, though they took place systematically in Alexandria after Hippocrates' death.[31] Anatomy and medicine flourished along with other forms of learning in Alexandria, especially under the Greek anatomists and physicians Erasistratus (304–250 BCE) and Herophilus (335–*c.*280 BCE).[32] Erasistratus is credited with describing the four ventricles in the brain that we recognize today, as well as the cerebrum that governed the whole nervous system and the cerebellum, seen as the site of the soul and intelligence.[33] Erasistratus also described sensory awareness as a product of psychic pneuma, a special substance endowed with the power to perform motor, sensory, and mental activities.[34] This theory evolved into a 'three cell theory' of brain function, in which each cerebral ventricle was the seat of a specific function, with a unique type of spirit and power to perform it.[35] The three-cell theory represents the earliest attempt to localize different mind functions in separate brain sites, and it held true across Byzantine, Arab, and Western medical theory at least until the Renaissance. Even before Ferrier's Goulstonian Lectures, then, theories of brain localization existed in another form.

The rise of the brain in Greek medicine was characterized by a broader philosophical interest in the location of the mind and the soul. Despite the heart's role in mediating between the soul and the passions, and in summoning the spirits to action, the brain allowed us to think and imagine, as well as to move and digest. Largely by dissecting animals Galen found that a freshly dissected cerebellum was hard to the touch and the cerebrum soft. This led him to conclude that the cerebrum was the recipient of sensations, that somehow it formed and imprinted memories into the fabric of the body, while the cerebellum controlled the muscles. There are several accounts of vivisection that detail Galen's deeply unpleasant experimentations on animals that severed the spinal cord and the nerves in animals at various places in order to see which parts of the body and its functions were affected.[36]

By the end of the fifteenth century, dissections of human brains became more common as a small cadre of French and Italian professors, inspired by the learning of the ancients, illustrated lectures from ancient texts.[37] The Italian physician Jacopo Berengario da Carpi (1460–1530) described the process of opening up a human head, comparing it to the chopping of an onion:

> If you should cut an onion through the middle, you could see and enumerate all the ... skins which circularly clothe the center of this onion. Likewise if you should cut the human head through the middle, you would first cut the hair, then the scalp, the muscular flesh (*galea aponeurotica*) and the pericranium, then the cranium and, in the interior, the *aura mater*, the *pia mater* and the brain, then again the pia, the aura mater, the *rete mirabile* and their foundation, the bone.[38]

Most anatomists dissected bodies simply to confirm the findings of the ancients, though Berengario produced the first anatomical text with detailed illustrations. *Anatomia Carpi* (1535) emphasized the importance of anatomical dissection rather than textual learning as well as using human cadavers, rather than animals.[39] Berengario claimed to have dissected hundreds of human bodies, and he also claimed he was first to deny the *rete mirable* (Latin for 'wonderful net'), a complex web of veins and arteries used to regulate temperature that Galen saw in animals and mistakenly extrapolated to humans.[40]

The sixteenth-century anatomist Vesalius later claimed that it was *he* who had discovered the non-existence of the *rete mirable*. Unlike his contemporaries, who classified ligaments, tendons, and aponeuroses (sheet-like tendinous expansions that connect muscles with the parts they move) as nerves, Vesalius classified a nerve as a mode for transmitting sensation and motion. Moreover, he rejected the claim that nerves originated in the heart, a belief that had been in place since Aristotle, and located them instead in the brain. In fact, Vesalius identified seven pairs of brain nerves and thirty pairs of spinal nerves, and the seventh book of the *Fabrica* is entirely devoted to the brain.[41] Until Vesalius, most work on the brain straddled the material and immaterial worlds; it suggested that the soul could work in and through the body, using for example Erasistratus' widely accepted idea that brain functions were carried out in the ventricles by the animal spirits. Vesalius rejected this belief, arguing that animals also had cerebral ventricles, though they certainly did not possess the rational soul of humans.

VITALISM AND THE MECHANISTS

These kinds of questions were polarized in debates between mechanists and vitalists, such as the English physician Thomas Willis' *Cerebri anatome* (*Anatomy of the Brain* (1664)) and the French philosopher René Descartes, whose *Passions of the Soul* (1649) Willis expanded on and critiqued.[42] Mechanists, like Descartes, believed that the actions of the body—from the heartbeat to the passions—could be explained through physics and chemistry, while the vitalists like Willis argued for a soul-like force that infused the body with motion and life.[43] Willis' interest in the brain stemmed from his concern for the location of the soul and his professed desire to 'unlock the secret places of Man's Mind and look into the living and breathing Chapel of the Deity'.[44] Willis concluded that the soul acted within the blood and the arteries rather than in the brain, a form of materialism that also existed in the English physician William Harvey's *Exercitatio anatomica de motu cordis et sanguinis in animalibus* (*On the Motion of the Heart and Blood*), published in 1628.[45]

Although Willis published *Cerebri anatome* less than forty years after Harvey published *Exercitatio anatomica de motu cordis*, the political climate had changed radically between the two books. That change is reflected in the relative statuses of the brain and the heart as anatomical organs and metaphorical objects. Harvey published his work fourteen years before the start of the English civil war; he described the heart as the body's centre, as its sun and its king, responsible for sending out spirits through the blood to the rest of the body politic.[46] By contrast, Willis wrote his treatise after the Restoration, when Charles II had taken the throne. Although he, like Harvey, stressed the importance of the fluids in conveying spirits around the body, he shifted responsibility from the heart to the brain. In Willis' work the brain was a 'kingdom', a 'chest', and a 'vault', both the 'chapel of the deity' and its ruler.[47] Nerves were 'silver and gold', spirits that acted as 'many distinct troops or companies of soldiers' and caused muscular movements 'like the explosion of gunpowder'. The reference to Charles II's right to rule over his land, albeit under the watchful eye of parliament and the Church of England, was made explicit.

In the seventeenth century all physiological explanations of brain and body interaction used fluids and animal spirits. In the eighteenth century, with the growing use of microscopy, a material alternative to the animal spirits (that often acted as the soul) emerged in the work of the Swiss doctor Albrecht von Haller. In *Primae lineae physiologiae* (*First Lines of Physiology*, 1747) and *Elementa physiologiae corporis humani* (*Elements of Physiology*, 1757–66), Haller disputed that there was a soul either in the ventricles of the brain, in the vital fluids or in the blood. Haller was a devout Christian and he did believe in the soul. Like Isaac Newton, however, he was influenced by the physical and natural laws that governed existence.[48] Haller introduced the concept of irritability to explain how muscles moved without a spiritual force. Even the heartbeat could be explained by reference to its internal structure.[49] The body could therefore be divided into parts that were 'irritable' and those that were 'sensitive', the latter having nerves to transmit pain.[50] Haller also differentiated the brain further, separating out the outer grey matter of the cerebral hemispheres and cerebellum and the white matter that he associated with sensory impressions. The seat of the mind was in the medulla, Haller concluded, the place where the spinal cord and cranial nerves originated.

Vitalists reacted against this mechanistic perspective. Robert Whytt, professor of medicine in Edinburgh, rejected Haller's views entirely, arguing that a non-material, 'sentient principle' must act through the brain and nerves to bring the body to life in the first place.[51] The debate between Haller and Whytt summed up a crucial and apparently intractable problem: was the body an automatic machine as distinct from the mind or was there some soul-like or animating principle at work? By the end of the century, the problem became more complex with the development of electricity, which would replace both Whytt's sentient principle and Haller's description of irritability and sensibility.[52] It would also help contribute to a whole new set of body-related metaphors of power and energy and waste.[53]

Electrical energy had been known since ancient times. As early as 43 CE, Scribonius Largus, court physician to the Roman emperor Claudius, used electrical currents to treat headaches and gout by applying electric torpedo fish to the affected regions.[54] Isaac Newton had suggested some electrical force might explain the way the mind and the body moved in unison, but it was not until the late eighteenth century that Luigi Galvani undertook a series of experiments that led him to conclude muscles retained electric power.[55] In a reformulation of animal spirits, Galvani viewed electrical energy as a non-material life force originating in the brain and flowing through the nerves to the muscles. It is said that Galvani's experiments on frogs meant he was nicknamed the 'frog's dancing master', but that is probably apocryphal.[56] After Galvani's death his nephew Giovanni Aldini performed a series of dramatic demonstrations that revitalized dead bodies, laying the foundations for experimental physiology as well as providing inspiration for Mary Shelley's *Frankenstein*.[57] Electricity provided a model to illustrate how brains could be 'galvanized' into action and send information through the nerves to the rest of the body like a telegraph system.[58]

MYSTERIES OF 'MIND'

If the origins and mysteries of the mind were traditionally explained in terms of a nervous force imbued with the soul, the rise of irritability and electricity allowed for a solely materialist perspective. With nineteenth-century neuroscientific concepts of localization, the material brain was seen as a series of 'little souls' that composed the 'general soul' of the entire body, before mind itself *became* soul.[59] Mirroring the process by which the heart was secularized and divested of any sacred power, human consciousness and thought were mapped onto the material brain. The nervous system, with its pathways and ganglions held the key to emotions, the personality, and the self.

Understanding how the brain functioned was a metaphysical and philosophical as well as a narrowly scientific concern. Could feelings like love, anger, joy, and hope be converted into physical activities or material movements along a nervous thread? Were the mind and its experiences physical, concrete entities that could be cognitively processed? Not all theorists believed in the reduction of emotion to physical processes, or of mind to matter. For those who did, it was not *whether* immaterial and psychological influences were felt in the material and physiological realms, but *how* they were felt. The role of cognition, of acquiring knowledge and understanding through thought, emotion, and the senses, was crucial to this process. Under the craniocentric model, 'normal' and 'pathological' emotions were linked to the material structure of the brain as well as its cognitive processes. The body might feel emotions (as manifested by a raised heartbeat or a cold sweat), but by the late nineteenth century the brain was seen to process those feelings in much the same way as the soul had done under humoralism. Reconceptualizing the brain as a cognitive centre (the term 'cognition' coming from the Latin 'to know' or 'to recognize') was essential to the rise of the mind sciences.

Surprisingly then, given its role in cognition, most physiological accounts of brain activity were unconcerned with emotions. From the Edinburgh-based work of Alexander Walter and Charles Bell to François Magendie's experimentations on the spinal roots of dogs, more research was focused on sensation and movement (and later, with the work of Broca and Wernicke, with localization in terms of language and movement).[60] This was one of the complaints made by the American physiologist William James in his classic paper 'What is an Emotion?' (1884). Exploring the body's influence on the brain, James rejected physiological theory that ignored 'the aesthetic sphere of the mind', its longings, pleasures, and pains, as well as its emotions.[61]

James suggested that the nervous system was predisposed to respond in particular ways, and that it was physical sensation rather than thoughts that triggered bodily experiences, from a raised heartbeat to a cold sweat. Only *after* those changes did the brain respond by recognizing an emotion as, say, 'fear'. Thus:

> Common sense says, we lose our fortune, are sorry and weep; we meet a bear, are frightened and run; we are insulted by a rival, are angry and strike. The hypothesis here to be defended says that this order of sequence is incorrect, that the one mental state is not immediately induced by the other, that the bodily manifestations must first be interposed between, and that the more rational statement is that we feel sorry because we cry, angry because

> we strike, afraid because we tremble, and not that we cry, strike, or tremble, because we are sorry, angry, or fearful, as the case may be. Without the bodily states following on the perception, the latter would be purely cognitive in form, pale, colourless, destitute of emotional warmth. We might then see the bear, and judge it best to run, receive the insult and deem it right to strike, but we could not actually feel afraid or angry.[62]

The American physiologist Walter Bradford Cannon famously disagreed with the 'James-Lange' perspective, so called because of similar contemporary arguments being made by the Danish neurologist Carl Georg Lange.[63] Cannon viewed the brain alone as the centre of all activity; emotional response came first, he argued, cognitively processed by the brain before being relayed to the body and causing the heart to race in anger or the hair to stand in fear.[64]

The work of both James and Cannon had a decidedly Darwinian influence, with the development of the brain being linked to evolutionary theory. In the 1880s and 1890s the English experimental physiologist John Hughlings Jackson also adapted Darwinian theory to build a conception of brain function as a series of layers, each of which was progressively more advanced than that which preceded it.[65] In ways that remain influential today (largely because they were adapted by Sigmund Freud in his schema, the *id*, *ego*, and *superego*), the emotions were associated with the most primitive functions of the brain.[66] Certain passions like anger were believed to be more 'primitive' than others, most notably love.[67] Interestingly, this hierarchizing of emotions was consistent with post-Enlightenment philosophy that prioritized 'social' over 'selfish' passions as well as the idea that controlling one's passions was an archetypal feature of modern, civilized society.[68] Inability to do so, as in the case of Phineas Gage, meant that one could be biologically, as well as socially, unfit for 'polite society'.

MATERIALIZING DIFFERENCE

In Austria, at about the same time as Galvani was working on his theory of animal electricity, Franz Joseph Gall (1758–1828) was developing a doctrine of skull shape called Phrenology. Associating distinct regions of the brain with particular motivations and emotions, phrenology reduced personality and intelligence to twenty-seven powers or functions that included traits like 'pride', 'cunning', and 'poetic talent'.[69] For Gall, the brain, and especially the cerebral cortex, was the 'source of all the feelings, ideas, affections and passions'.[70] Gall did not focus on specific emotions, but on such general sentiments as 'self-esteem' and 'benevolence', which is partly why phrenology was debunked as a pseudo-science. Also unhelpful was Gall's identification of the brain as a muscle, and his erroneous belief that the brain determined the shape of the skull.[71] Nevertheless, Gall's work provided a theory of localization that, as we have seen in the work of Ferrier, helped to explain what happened to Phineas Gage.

Nineteenth-century brain science fixed intellectual and moral deficiencies in the brain just as phrenology had fixed them in the skull (and Cesare Lombroso's work would fix them in a person's 'atavistic' physical defects).[72] Prioritizing the brain and its emotions as the centre of knowledge and the self had important social and political impact: women and non-European men were all found to have less developed brains than white, middle-class European men. In discussing brain size, Broca found a 'remarkable relationship' between intelligence and brain volume. Since men had larger brains they were necessarily more intelligent: women's brains were lighter by at least five ounces.[73] Perceived differences in their brains, moreover, meant that some women were more naturally 'feminine' than others. The Edinburgh surgeon John Bell wrote about an indulgent mother with a highly developed 'philoprogenitiveness' region that indicated her love of children, and contrasted it with his study of the brains of thirty women with defects in that region who had also been found guilty of infanticide.[74] Phrenologists had similarly found there was less 'vigour' in female intellect and narrower faculties of reflection, that women were guided by feelings and not intellect, were more timid and cautious, and had smaller areas of their brain dedicated to sexual love.[75]

Despite its limitations, phrenology had the same powerful, visual, and mass appeal that craniometry and craniology would have in the work of the 'skull collectors', those nineteenth-century cranial gatherers who amassed and compared groups of skulls to 'prove' the dominance of some races over others.[76] They argued that if women were lower in intelligence than men, then the same was true of black men as compared to whites. 'Men of the black races', after all, had brains 'scarcely heavier than that of white women', according to one of Broca's colleagues.[77] The American naturalist Samuel George Morton's ideas about brain difference underpinned early physical anthropology and the development of American craniology. Morton collected more than one thousand human skulls to show hierarchical differences between the races, sorting his boiled-down crania according to such fictional racial groups as Caucasian, Mongolian, American, Malay, and Ethiopian. The intent was to show the supremacy of Caucasian skull size, and therefore brains, in relation to other ethnic groups.[78]

Some of these ideas stay with us, including the gendering of brains. A 2014 study argued that men have larger total brain volume than women. It also specified sex differences in volume and tissue density in the amygdala and hippocampus, the regions connected to emotion and memory, and in the insula, the 'wellspring of social emotions', a claim that reinforces traditional ideas of women as natural nurturers and caregivers.[79] Other neuroscientific studies cite the slow evolution of the adult brain as the reason why teenage girls are often obsessed with looks and why women supposedly develop 'mommy brain', a state where women are supposedly more forgetful and easily distracted after pregnancy.[80]

Today, popular science books detail such presumed material brain differences as 'proof' that women are more risk-averse than men and that men are more competitive and career-driven. Contested biological absolutes are used to construct whole narratives about life, work, and pleasure.[81] Naturalizing gender difference is problematic, even or especially when couched in post-feminist language about women being emotionally superior to men. This is recognized by other neuroscientists who debunk this notion, citing the role of environmental factors in making a difference. The circuitry of the brain is not as influential as the steady drip of gender stereotyping.[82] And the theory of plasticity means that a woman's brain might ultimately become wired for multi-tasking or empathetic response, for instance, not because of any innate factors, but because social expectations demand she use that part of her brain more often. Just as repetitive use increases the size of the bicep, specialized parts of the brain develop with use.

THE SURVIVAL OF THE IMMATERIAL SOUL

This chapter has provided something of a historical overview of the anatomizing of the brain and its functions, especially in relation to mind. I want to conclude this chapter with further consideration of the soul. Today we presume that the spiritual soul has no place in the secular brain. In Western medicine its influence has diminished with the rise of such disciplines as experimental psychology and psychiatry.[83] The modern medical paradigm takes it for granted that we are entirely psychochemical and biological creatures, that we equal the sum of our material processes. We think, feel, believe, and remember, and though we might not yet be able to pin all of those phenomena down to physical processes, they are presumed to take place in the physical brain and not in any ethereal, immaterial place. The 'soul' is thus mere metaphor; as one scholar has put it:

> The mind or the soul is the brain. Or better: Consciousness, cognition, and volition are perfectly natural capacities of fully embodied creatures engaged in complex commerce with the natural and social environments. Humans possess no special capacities, no extra ingredients, that could conceivably do the work of the mind, the soul, or free will as traditionally conceived.[84]

Yet the emergence of a secular model of the mind should not blind us to the fact that theological or metaphysical explanations do survive; they are simply not granted intellectual or academic status in the West.[85] Moreover, despite the rise of the brain as the organ of the self par excellence, non-Western and non-orthodox medical traditions do prioritize other organs, including the heart. In any case, there is always a division between theory and practice; the fact that Gage's emotional changes were linked to material damage in the brain did not mean that he was presumed to be soul-less.

Consistent cultural belief in the soul, or in the immaterial mind located in the brain, is one of the main reasons why head or brain transplants cause such anxiety. Like heart transplants, they strike at the centre of ideas about the integrity and sanctity of 'the self'. Since the 1960s films like *The Brain That Wouldn't Die* (1962)—in which a doctor keeps his girlfriend's head alive after a car crash—highlight social anxieties about the role of science in 'playing God' (see Fig. 7).[86] Brain transplants are not currently a scientific possibility, but every so often they seem to be.[87] In 1971 the American neurosurgeon Robert J. White successfully transplanted the head from one

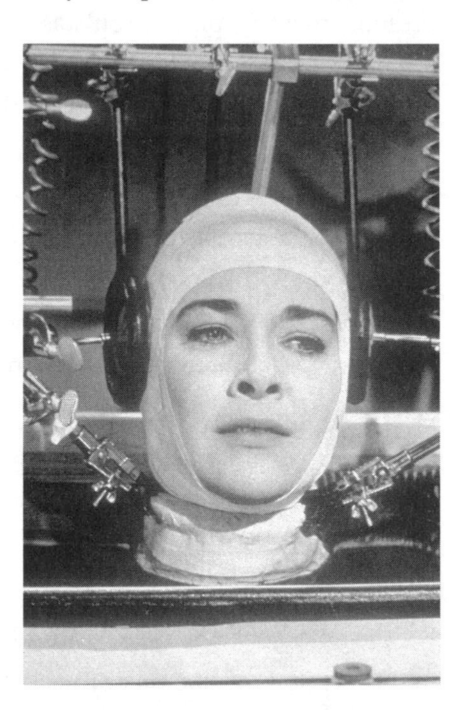

Fig. 7. Still from the American science-fiction horror film *The Brain That Wouldn't Die*, 1962, starring Virginia Leith (pictured) and directed by Joseph Green.

living rhesus macaque monkey onto the body of another and published his results alongside a series of hasty sketches.[88] Because the surgery involved severing the spine at the neck, the monkey was paralysed from the neck down but it could still hear, smell, taste, eat, and follow objects with its eyes. After nine days, and presumably in horrific pain and distress, the monkey died as a result of immunorejection.[89] The ethical and moral implications of White's work, in particular his 'barbaric' practice of vivisection and head transplantation on more than one hundred rhesus monkeys, earned him the title 'Dr Butcher' amongst animal rights' protestors.[90] Nevertheless, buoyed by his apparent success with one form of primates, White planned to perform the same operation on humans in the 1990s, routinely practising on corpses. In an interview with the BBC Today programme, White presented the possibility of head transplants as the next obvious step in the history of transplantation, emphasizing its role in treating people whose bodies were paralysed but their brains completely healthy.[91]

Reactions against White's work were widespread, and not only from animal rights activists. The neurobiologist Professor Stephen Rose has condemned White's proposals, describing them as 'medical technology run completely mad . . . scientifically misleading, technically irrelevant and scientifically irrelevant, and apart from anything else a grotesque breach of any ethical consideration'.[92] If such a procedure were fully realized in human beings, the issue of who a person would 'be' would become immensely complicated. Ethical considerations have been a problem at each stage of the history of transplantation, including the heart.[93] Those questions are obviously magnified because it is the brain that is now identified as the repository of our feelings, emotions, and memories as well as the basis of thought; not only in a material sense, but also an immaterial one.

The philosophical problem of the mind clearly continues above and beyond any physical explanation about the brain.[94] This can be seen by a recent discussion of modern anatomical dissection by medical students. The process of anatomical dissection presumes objectivity and distance, but that is not the case; medical students are no less influenced by cultural beliefs about the soul than anyone else. Some students reported that it felt 'brutal' to invade the 'personality' of the dead by removing the human brain from the skull—perhaps partly because of the proximity of the saw to the human face, 'which made it obvious they were cutting up a person'; dissecting the brain away from the body made the process easier because 'it was kind of an organ and it was easier to view the brain as an organ and dissect it from there'.[95]

Other students, however, associated the brain with something above and beyond the physical realm, even once it was separate from the body. The brain was figured as 'the most important part of you' and 'the thing that made you who you are', which is a pretty unsurprising response in our neurocentric age. But the brain was also referred to as the 'whole soul' of a person: 'you're holding or cutting what made that person them, what made them special and unique and I think there's always... something a bit more sacred and special about a brain than anything else'. Of course this same cohort of students made similar assertions about the heart, but that only makes their observations more intriguing. The brain was compared to the 'motherboard of a computer' and to a 'mass of wiring', as well as a 'tiny little bit of pudding' or a 'sponge', though it was also felt to possess qualities above and beyond its physical structure.[96] To cut up a brain, it was suggested, one needs to retain 'a certain separation there between this as a person and this as a specimen'. A similar observation has been made by the neurosurgeon Henry Marsh in relation to brain surgery:

> I look down my operating microscope, feeling my way down through the soft, white substance of the brain, searching for the tumour. The idea that my sucker is moving through thought itself, through emotion and reason, that memories, dreams and reflections should consist of jelly, is simply too strange to understand. All I can see in front of me is matter.[97]

Even neuroscientists cannot excise the soul from the brain as they might a tumour. Whatever the immaterial mind consists of—consciousness, self-hood, identity, memories, feelings, and thought—it remains elusive. The 'mind' that was once explained as humours working through the soul or animal spirits is today reduced to a heady mix of chemical and electrical processes and hormones that wind their way through one hundred billion neurones.[98] The mind is in the brain but it is also of the body and its processes. The metaphors we use have evolved to acknowledge both the brain's complexity and its communication with the rest of the body: Willis' 'silver and gold' nerves that sent 'explosions' of gunpowder through the brain have become electrical impulses and computational networks, with more emphasis than ever on the way the mind communicates with all the body's nerves and fibres.[99]

A more significant challenge to the primacy of the mind *exclusively* as brain is that parts of the body seem to communicate back. We have seen how the brain gradually 'won out' over the heart, at least in terms of the location of mind. The brain has retained its metaphorical significance too, seeming to spawn a 'little brain' in the heart. And it mediates between the mind and the rest of the physical body, which includes not only our internal organs, but also our most visibly expansive external 'organ': the skin. It is to the skin that we must now turn, that communicative part of our selves *par excellence* on which is writ our thoughts and fears, our emotions and affectations.

6

From Excrement to Boundary

Touching on the Skin

> You never really understand a person until you consider things from his point of view until . . . you climb into his skin and walk around in it.
>
> (Atticus Finch in *To Kill a Mockingbird*[1])

On 1 April 1814, Charles Uncle, a fourteen-year-old boy living in London, was admitted to St Bartholemew's Hospital suffering from a distressing skin complaint—a series of raised welts and 'tubercules' that had appeared all over his face. Charles, a pale child with brown hair and dark eyes, met William Lawrence, a professor of anatomy and Fellow of the Royal College of Surgeons, and Henry Southey, a physician at the Middlesex Hospital. The following February Lawrence and Southey reported on that meeting to the *Medico-Chirurgical Transactions*, the publication of the Royal Chirurgical [Surgical] Society of London.[2] That report took the form of a case study from which we can learn much about Charles and his life, as well as the diagnosis of skin disease. His experience, moreover, is a useful introduction to the significance of skin in history; it also hints at the pain endured by those who lived with disfiguring and often untreatable diseases.

Skin arguably reveals more about ourselves, as individuals and in relation to others, than any other part of the human body, including how and where we live as well as our inner experiences. It communicates those experiences to others through physical expressions of emotion like blushing, paling, goose pimples, and sweating. The skin is the outer covering of all vertebrates and the largest human organ.[3] It envelops the muscles, bones, ligaments, and internal organs, marking out the physical boundaries of the individual and protecting us from both pathogens and water loss. The role of the skin in veiling and hiding the body's secrets was made manifest in Renaissance paintings of flayed skin in the legend of Saint Bartholemew, the apostle who was hanged upside down and flayed alive, and in anatomical tables and illustrated accounts of human anatomy.[4] Following Galen, the Arab polymath Ibn-Sīnā arranged the body in the following order: bones, veins, arteries, nerves, internal organs, and finally the skin.[5] In these works the skin was the first layer that a reader peeled back to examine the innermost secrets of the body, a series of flaps and cut out panels that invited the reader's participation and engagement with the world of flesh.

The skin mediates our engagement with both the social and the environmental worlds. Its nerve endings respond to heat and cold, touch and pressure, vibration and injury, alerting us to the nature of the environment as well as to danger. Skin regulates our temperature through sweat glands and blood vessels, enabling us to adapt to different conditions and ways of life. Unlike that of other animals, human skin is usually without much hair, aside for the scalp, groin, armpits, and male chin. Unlike many animals, and more than many mammals, the human body also sweats. Skin comes in many colours, from almost black to nearly white: a natural gradient that is related to the intensity of the UV rays that fall at the different latitudes on the earth's surface.[6] The skin itself is a surface for decoration, as the individual uses his or her skin to fit into, or sit outside, social conventions.

Being free of blemishes, of scars, stretch marks, wrinkles, of all the signs and signifiers of our lives and our experiences as human beings, has become a modern-day ideal. This is all the more important in an age when so much is made public and widescreen and immediate through multimedia technologies. As James Joyce put it even before the onset of Twitter and the age of the selfie: 'it is almost as though modern man has an epidermis rather than a soul'.[7] Our quest for perfection, for youthfulness and smoothness, is manifested in the growth of the cosmetic surgery industry and in skin-care sales estimated at more than 43 million dollars a year in the United States alone. Europe and the US are the biggest consumers, united in the pursuit of perfection and the avoidance of the visible signs of ageing. Dermatological conditions are also one of the top fifteen groups of medical conditions in terms of prevalence and medical spending in the US. Approximately one-third of American people were believed to have some form of skin disease between 1987 to 2000. This estimate covers conditions as diverse as acne and carcinoma.[8]

It was during Charles Uncle's time that many of these diseases were first identified. Medical understandings of skin were transformed in the nineteenth century, along with many emerging specialisms. Dermatology as a distinct clinical field moved away from earlier definitions of the skin as a waste product excreted by the body (which was the position of humoral medicine), towards viewing skin as both an anatomical organ and a source of sensory experiences. Classifying diseases of the skin, based on the appearance of patients like Charles Uncle, was part of the process by which the abnormal was categorized and compared with the normal. The healthy ideal was and is unmarked skin, free of the taint of disease. This chapter explores the history of the human skin as a sensory object capable of being marked by disease, as well as a blank canvas on which we mark, deliberately or otherwise, feelings, emotions, and social identities. Beginning with the case of Charles Uncle, it considers some of the material, moral, and even spiritual meanings placed on the skin.

CHARLES UNCLE AND THE SCIENCE OF SKIN

We can learn much about Charles Uncle through the testimony of his physician in his written account to the Royal Chirurgical Society, though sadly little of Charles' own subjective experience. This is a lamentable lack in many medical records dealing with the experiences of ordinary men and women. Lawrence reported that Charles had been born in America to English parents. They married young before moving to America, where Charles and his siblings were born. Those siblings, a girl and a boy, were sent to England when they were young, and neither had experienced any form of skin disease. The girl died, however, of 'consumption' (tuberculosis), shortly after contracting measles when she was just sixteen years old. While Charles was still young his father also died, and he was sent to New Providence in the Bahamas. His 'father-in-law', presumably his step-father, as Charles was still a young boy, made Charles work in the fields in the heat, exposed to the sun and all weathers, and he was given the same 'coarse' food that was 'given to negroes'. Yet he remained in good health until he boarded a ship to return to England. Charles was forced to work on the ship to pay for his passage and on more than one occasion he got very wet and acquired a terrible cold. Charles 'felt himself very ill and drowsy but his appetite did not fail'. His face and head started to swell, and when the swelling subsided large tubercules appeared all over his ears and face and his limbs grew stiff. By the time he was admitted to Bartholemew's and placed under the care of Lawrence and Southey, the disease had spread to all his limbs.

An illustration of Charles' diseased face appeared in the work of the physician Thomas Bateman, a Yorkshireman who was working in London as a pioneer in the field of dermatology (see Fig. 8). Bateman relied on the efforts of his associates, like Lawrence, to bring him new and interesting cases for his *Delineations of Cutaneous Diseases*, which completed the work of his erstwhile colleague and mentor Robert Willan, about whom more will be said below. Charles' portrait appeared under a discussion of 'tuberculosis elephantiasis', one of the diseases in his newly developed classificatory system. Bateman was pleased with the portrait, describing it as an 'exact' likeness 'after the tubercules had been rendered a little smoother than they originally appeared by the application of poultices'.[9]

The trajectory of the disease was slow. First Charles had noticed a series of 'flattened tubercules' breaking out on his skin. They were 'skin

Fig. 8. Charles Uncle, a skin disease patient, whose condition was used as an example of tuberculosis elephantiasis by Thomas Bateman in 1817.

coloured; not larger than 'half a small pea at first', but gradually increasing to a 'much more considerable size'. Soon they changed colour, becoming red, and acquiring 'in some instances a deep tint of that colour with a rather livid cast':

> In some parts they remained in this state: in others an abundance of white and small scales was formed. Some of the tubercules cracked and ulcerated; but the ulcerations were not in general deep or extensive: they furnished a matter which concreted into hard crusts, and caused the dressings to stick very firmly.[10]

Except for a number of 'fissures and alterations', and the painful removal of dressings that were stuck to the lesions, Lawrence reported that Charles did not suffer unduly. At the time of his admission, though, his 'ears, forehead, eyebrows and eyelids, and indeed the whole face were completely occupied by the disease'. His nose had flattened, his ears had become misshapen, his lips and cheeks swollen, and his eyebrows had fallen out though the hair on his head remained in place. The tubercules even extended into the poor child's throat, though his eating seemed to be unimpeded. His limbs, especially his feet, were swelled and covered in tubercules, and his scrotum had 'shrivelled and seemed empty...The testes could with difficulty be felt; they were soft and about the size of small horse beans.'

Soon after Charles had been hospitalized, his symptoms grew worse. More tubercules began to appear, causing him dreadful discomfort if he tried to speak or to swallow. The only relief he found was through ointments and emollients that loosened and softened the crusts on the tubercules. He was fed medicines of mercury and arsenic, commonly used for skin diseases related to syphilis, for instance, but these made him feel much worse and that in turn seemed to 'aggravate the complaint'. Instead the doctors tried sulphuric acid, which seemed to bring some relief. In general, however, all medicines were ineffective, though he managed to eat a reasonable diet of meat, porter (a dark, malt beer), and wine.

By 2 February 1815 the skin disease seemed to have abated and Charles was discharged. During his stay in hospital all existing tubercules had healed themselves and no new ones appeared. Doctors were puzzled as to the sudden cure. They noted that Charles had some permanent disfigurement as a result of the tubercules, especially on his ears and lips, and that his skin was scarred. However, it soon became apparent that the boy had developed some 'internal' problems that necessitated his readmittance to hospital; in particular a pronounced cough that was common to tuberculosis cases. Charles was short of breath and weak, and skinny for his age. When his cough subsided he was sent to live with his relatives in Devonshire, presumably because the air would have been much cleaner than in London. On 9 May Charles wrote a letter to Williams, an extract of which was produced in the journal, which gives an unusual insight into Charles' own words, though of course they would have been mediated by letter-writing conventions of the time, and the expected deference of a patient to his physician:

> My bodily health is much improved with respect to strength and eyesight; but I have within a week broken out in three or four places about my face, which I think is merely change of climate. I am, according to your advice, placed at a farmhouse, where I am comfortable. I amuse myself with shooting and fishing and reading.[11]

Another letter received from the boy's mother at the end of June confirmed the bad news that the condition of his face had again become poor and the tubercules had returned. Charles' health seems to have fluctuated over the following months, during which time his brother died of tuberculosis. We do not know what happened to Charles after August. The only reason he comes to light in the historical record, as a relatively poor boy, albeit one who hunted and read, is because he acted as a case study for the rise of dermatology, and was given charitable treatment at one of the most eminent London hospitals.[12] Along with that of a woman, a prostitute known only as 'Miss N', who was admitted alongside Charles and for the same disease, Charles' case caused quite a stir among the medical profession—and beyond. According to one source the poet Percy Bysshe Shelley was terrified he might have caught syphilis (often connected to elephantiasis) through contact with the same 'Miss N'. This came to light only because Charles and Shelley apparently shared a physician.[13]

Willan and Bateman were not the first doctors to describe skin disease. As early as 1025 CE, the Islamic physician Ibn-Sīnā described treatments for a variety of skin conditions, including cancer, in *The Canon of Medicine*. In Ibn-Sīnā's work the skin indicated internal disorders, as 'sometimes the "crude" humours are situated in the flesh'.[14] It was common to regard the skin as excrement in the ancient world, produced by the body as a discarded element rather than as a boundary of the person or the body. Aristotle viewed the skin as an after-effect, a hardening and drying of the external body, in much the same way as scum might form on a boiled surface. His view made perfect sense in a culture that focused on the fluids, rather than the solids and fibres of the body, and on a symbiotic relationship between self and world. Thus in his *Book of Treasures*, the Syrian philosopher Job of Edessa wrote of skin as coming into existence externally, almost like skin hardening on a bowl of rice pudding:

> When the humidity of the outside portion met the air, the latter destroyed the thinness which it possessed, and it thickened; and as a result its parts came together, solidified, and became skin. The same thing happens when we cook grains of wheat, or other things; after they have dissolved and become chyle, if we leave them a short time exposed to air in a vessel or plate, the humidity rises above them and forms a skin in an outside position, in such a way that we can take it with our hands.[15]

In the humoral tradition skin had an important role in the maintenance of health. It was less a barrier through which the as yet undiscovered 'germ' could pass than an agent in the regulation of temperature and fluids.[16] The excretion of moisture through the skin by sweating or even by weeping helped to explain how the fluids moved from one part of the body to another. It also suggested that fluids could move from one organ to another just as it progressed through the canals and conduits of the human body. A swelling or boil similarly showed that there was some matter within the body that was trying to get out and was pushing itself towards the surface. In this context, Charles' skin disease would originally have been regarded as a symptom of some internal imbalance, rather than a specific disease entity; treatment would have centred on purging, bleeding, and the maintenance of the 'non-naturals', those factors that are extrinsic to the body–including sleep, diet, exercise, air, excretions, and the passions–but which were crucial to maintaining health.[17]

What was distinct about Willan's and Bateman's work was not so much their treatment of skin diseases–like many other physicians they made recourse to humoral methods as well as to mercurial treatments–but their careful classification of those diseases according to distinct visible signs. In earlier periods, skin diseases were based on more general symptomology rather than appearance. Willan was working as a physician at the Carey Street Public Dispensary when he met and worked alongside Thomas Bateman, drawing his conclusions about dermatological conditions from patients he treated, such as bakers, for whom psoriasis diffusa on their arms and hands seemed to be an occupational hazard.[18] Willan developed a taxonomic classification for diseases like impetigo, lupus, psoriasis, ichthyosis, and others that was based on their morphology, or appearance, and influenced by the systematic principles introduced by Carl Linnaeus, the Swedish botanist who named specimens by binomial nomenclature: i.e. a two-part name that gives first the *genus* and then the *species*.[19] Willan provided both classification and clinical description in the first volume of his work *On Cutaneous Diseases* (1808), which not only categorized skin diseases based on the physical shape, size, and colour of the marks on the body, but also provided detailed illustrations. When Willan died in 1812, Bateman continued his work, publishing first *A Practical Synopsis of Cutaneous Diseases, According to the Arrangement of Dr Willan* (1812),[20] and subsequently his atlas, *Delineations of Cutaneous Disease,* which featured the portrait of Charles Uncle.

The emergence of dermatology can be seen as part of a longer scientific trajectory that discerned differences based on physical appearances in healthy as well as pathological conditions. Central to the reinvention of the body as a scientific object was a change in the understanding of the nature and function of the skin, of its perceived ability to feel pain, the role of colour, and the skin's ability to communicate human emotions like fear and anger and love. Yet not all humans were believed to 'feel' alike and sensory experiences differed; comparisons between black and white skins were already well established in scientific culture before the time of Charles Uncle.

At the same time as European people were comparing themselves to non-Europeans as a result of imperial expansion, racial identities were evolving into whites and non-whites. The complexion of Elizabeth I, the virgin queen, was invested with a series of interpretations about purity and racial hierarchy even before the emergence of 'scientific racism'—the nineteenth-century use of pseudo-scientific techniques to construct and justify concepts of race and racial hierarchies.[21] Just as blackface performances by actors on the Shakespearean stage fused dark skin with animalism and savagery, whiteface cosmetic practices promoted British whiteness as an indicator of moral, and later intellectual superiority.[22]

VISIBLE DIFFERENCE: THE INVENTION OF RACE

'Race' has been used to indicate cultural and ethnic differences, geographical and historical, religious, and even linguistic ones. It has not always signified colour. In 1694, for instance, the *Dictionnaire de l'Académie Française* defined 'race' as 'lignée, lignage, extraction' in relation to families or beasts. The same definition is given in the *Encyclopédie* (vol. 13, 1765), where 'race' is linked to the idea of a 'noble race' or family. Not until the nineteenth century, with the sixth edition of the *Dictionnaire* (1835) was race defined as: 'Une multitude d'hommes qui sont originaires du même pays, et se ressemblent par les traits du visage, par la conformation extérieure' (a multitude of men who originate from the same country, and resemble each other by facial features and by exterior conformity).[23] Similar English dictionary definitions came even later, with graduations of skin colour between black and white attaching themselves to existing prejudices.

In *Black Face, Maligned Race* (1987) the literary critic Anthony Gerard Barthelemy showed how the ancient association of blackness with evil and sin was well entrenched in the Christian tradition by Shakespeare's time.[24] Certainly in *Othello*, blackness is identified with sexual power, passion, and violence.[25] Along with an allegorical reading of black skin as linked to the devil was the Christian theory of Africans as the descendants of Noah's sinful son, Ham. The double association of black people with bestiality, moreover, as well as sinfulness, dates back to the sixteenth century and the chronicler of John Lok's voyage to Guinea:

> It is to be understood that the people which now inhabite the regions of the coast of Guinea, and the middle parts of Africa, as Libya the inner, and Nubia, with diverse other great & large regions about the same, were in old times called Æthiopes and Nigritae, which we now call Moores, Moorens or Negroes, a people of beastly living, without a God, lawe, religion or common wealth, and so scorched and vexed with the heat of the sunne, that in many cases they curse it when it rises.[26]

One of the first attempts to define race was made by the French physician and traveller François Bernier, whose *Nouvelle division de la terre par les différents espèces ou races qui l'habitent* (*New division of Earth by the different species or races which inhabit it*), was published in 1684. Bernier's work separated Native Americans, North Africans, and South Africans on the basis of skin colour.[27] Others followed suit, including Linnaeus, who divided humans into continental varieties of Europanus (the white race), Asiaticus (the yellow race), Americanus (the red race), and Africanus (the black race). Linnaeus also borrowed from Galenic theory in order to associate each race with a different humour: sanguine, melancholic, choleric, and phlegmatic.[28] In 1775 the German physician, naturalist, and anthropologist Johann Friedrich Blumenbach followed Linnaeus by dividing human beings into races: the Caucasian or white race, the Mongolian or yellow race, the Malayan or brown race, and the Ethiopian or black race. Blumenbach's thesis was 'degenerative'; he argued that Adam and Eve were Caucasian and that other races subsequently emerged as a result of environmental factors such as the sun and poor diet.[29] The work of Linnaeus and Blumenbach and others, most notably the French naturalist Georges-Louis Leclerc, Comte de Buffon, effectively entrenched established ideas about human difference in civility, behaviour, and appearance into a distinct and coherent system of racial difference.[30]

With European colonization, many different political and cultural traditions were put under the microscope and classified—no less than the diseases of the skin—according to observable and hierarchized differences.[31] The growth of the Atlantic slave trade provided further incentive to ascribe personality types, intellect, emotional types, and behaviours to skin tone differences.[32] Thus the German philosopher Christoph Meiners (1747–1810), a 'polygenist' who believed each race had a different origin, and an early practitioner of scientific racism, defined black people as immoral and animalistic:

> The more intelligent and noble people are by nature, the more adaptable, sensitive, delicate, and soft is their body; on the other hand, the less they possess the capacity and disposition towards virtue, the more they lack adaptability; and not only that, but the less sensitive are their bodies, the more can they tolerate extreme pain or the rapid alteration of heat and cold; when they are exposed to illnesses, the more rapid their recovery from wounds that would be fatal for more sensitive peoples, and the more they can partake of the worst and most indigestible foods... without noticeable ill effects.[33]

Meiners claimed that black people felt less pain and experienced fewer emotions than white people, which resulted from their having thicker skin and less sensitivity than the cultured, more superior whites. While blackness was associated with skin insensitivity and animalism, whiteness signalled moral and sexual purity, especially in the tradition of British portraiture.[34] This trend increased as a response to social and political anxieties about the circulation and potential mixing of racial blood, and the assertion of a particular kind of British nationalism.

Whiteness was as important as blackness in defining race and status. In both African and Asian cultures, skin whitening has been associated with beauty for centuries.[35] While its history can be traced to the associations of virginity and purity found on the skin of Elizabeth I, there has been a disproportionate use of whitening treatments among people of African descent.[36] From the expansion of the British Empire in the Shakespearean age to the development of modern consumer society, it has been called the 'White Man's burden' to educate and civilize and lighten black skins, an association consistently made between skin colour, morality, cleanliness, and whiteness. As a *Pear's Soap* advertisement explained in 1899:

> The first step towards lightening The White Man's Burden is through teaching the virtues of cleanliness. Pear's Soap is a potent factor in brightening the dark corners of the earth as civilization advances, while amongst the cultured of all nations it holds the highest place—it is the ideal toilet soap.[37]

The ability of white skin to show 'feeling' was also politically charged, especially when linked to women's bodies and to virtue.[38] In the nineteenth century, under evolutionary physiology, blushing was attributed to an overactive sympathetic nervous system. In *The Expression of the Emotions in Man and Animals*, Charles Darwin wrote of blushing as 'the most peculiar and most human of all expressions'.[39] He related blushing to self-consciousness, as the 'mental states which induce blushing... consist of shyness, shame and modesty; the essential element in all being self-attention... It is not the simple act of reflecting on our own appearance, but the concern for what others think of us that excites a blush.' Today facial blushing is related to a number of factors in combination, including an extensive network of veins close to the surface and the relatively thin skin on the face. There is still relatively little known about blushing as a physiological and *psychological* phenomenon, but excessive blushing is a diagnosed medical 'problem' or 'disorder' for many people, especially those who experience a form of social phobia. There is even a term for a fear of blushing, erythrophobia, which literally means a 'fear of redness'.

Nineteenth-century scientists worried about blushing in women as well as men, in the old and the young, and in differently defined racial groups because of what it might tell them about the relationship between body and mind. There was 'no subject more interesting either to the physiologist or general enquirer', the British surgeon Thomas Burgess argued, 'than that which embraces a consideration of the involuntary acts of the mind upon the vital organs and their several functions'. According to Burgess' 1839 treatise, blushing was one of the most illustrative examples of this process.[40] While many writers—like the Scottish surgeon and anatomist Charles Bell—rejected the idea that black people could blush ('I can hardly believe that a blush may be seen in a Negro'), Burgess disagreed.[41] From a survey of the work of contemporary explorers Burgess concluded that people of African descent probably had the physiological capacity to blush, though it was not as developed in them as it was in the Caucasian races. Following Blumenbach's division of the human species into five distinct groups, Burgess opined that white skin 'with a fair rosy tint', possessed the highest 'moral feelings and intellectual powers', and that those of African descent might not blush for the same reasons as their white counterparts:

> We shall find that man, as he progresses from the savage state, in which he obeys or follows the dictates of nature, to that of civilization, wherein he observes the rules of art, advances *pari passu* [side by side] in the vices of its refinement. Who ever heard of an African savage blushing from morbid sensibility?'[42]

Darwin agreed that the vessels of the face were 'filled with blood, from the emotion of shame, in almost all the races of man, though in the very dark races no distinct colour change of colour can be perceived'.[43] Blushing was universal then, though the perception of the observer, and the ways the blush might be read, varied according to both its visible intensity and presumptions about ethnicity.[44] Despite this acknowledgement the prevailing belief that blushing was a fair-skinned phenomenon, did little to weaken the idea that black people were less moral and sensitive than their white counterparts.[45]

It is clear from these accounts that the blush has been racialized, which has implications for modern social and medical understandings of black and white skin. Scientific medicine in the nineteenth century ignored linguistic and cultural conventions about blushing, and it arguably still does. Leading Ghanaian scientist Felix Konotey-Ahulu has shown that African dermatologists who use English textbooks have to be cautious of white-skewed diagnostic use of the blush, as well as the differences in language between English and Ghanaian use of the term itself. Phrases like 'scarlet with rage', for instance, are 'untranslatable into many African languages, just as the Ghanaian term "his stomach swelled to bursting point" would not easily translate into English as signifying 'he was very, very angry'.[46] Anger was an emotion that Victorian writers perceived on the faces of black and white people alike; an extreme and uncivil emotion that also fitted in with ideas about the moral 'baseness' of criminals and savages.[47] From the nineteenth century onwards a series of physiological experiments sought to demonstrate not only that there was a science of emotions based on visible changes on the skin's surface, but also that there were racial differences in the skins and emotional expressions of black and white people.[48]

EMOTIONAL EXPRESSION

Emotional communication through the skin takes many forms. Fear makes the hair follicles stand on end, producing 'goose pimples', a phenomenon seen in other mammals beside humans, from cats to porcupines. It is not only fear that makes the base of the hairs stand on end, but also other emotions including fear, pleasure, and sexual arousal. The skin can also change colour and texture when a person is in shock or suddenly receives bad news. In the humoral tradition changes in the skin's appearance were explained by the movement of the blood and spirits: in fear, for instance, they rushed away from the fearful object, withdrawing towards the heart and making the skin deathly pale. In anger and passion, conversely, the spirits rushed towards the edges of the body, bringing a flush of blood to the skin.[49]

The emotional expressions found on skin began to be more carefully measured during the nineteenth century, in conjunction with all physiological and psychological processes. The historian Otniel Dror has shown how physiologists in the early twentieth century produced a range of objective devices by which feelings could be read in humans and in animals.[50] One of the ways emotions were read through the skin was through Galvanic Skin Response (GSR), which measured changes in the electrical resistance of the skin associated with emotion, attention, and stress.[51] Measuring emotional response through the skin by a series of bioelectrical experiments was central to the work of psychoanalysts like Wilhelm Reich, who prioritized the language of the body.[52] It was the Russian physiologist Ivan Romanovich Tarkhanov (or Tarchanoff) who first identified GSR in 1890 as an autonomic nervous system response based on physiological arousal to a stimuli—such as a word prompt—that might cause a change in bodily resistance.[53] Today, the principles of GSR have been widely popularized through the use of the 'lie detector', which presumes, like the rise in interest in body language more generally, that the skin cannot lie.[54]

Less well studied is the racialized history of objective measurement. In the 1950s it was argued that black skin demonstrated 'significant difference' in galvanic response based on experiments between 'one Negro and one white' subject.[55] In the 1970s, too, research was conducted to see whether there were 'racial differences' in the skin's resistance in GSR experiments—based on '12 Negro and 12 Caucasian' subjects.[56] More recently research has focused on the stereotypes associated with the facial expressions of men and women according to their different ethnic groups, and to the ways in which racial prejudice on the part of the subject and the examiner might affect perceptions of emotion.[57]

Nevertheless, investigation into the physiological differences between races continues to attract attention of scientists and the public alike, as does the popular adage that 'black don't crack', implying that black skin ages better than white.[58] In a series of 1990s experiments, biophysical differences between Hispanics and Blacks and Whites have been measured using a range of non-invasive methods to show that 'marked differences between races' exist in terms of elasticity, water content, and sun-damage prevention.[59] More recently racial differences in skin properties have been explored to explain differences between ethnic experiences of dermatological disorders, including hydration and lipid content and elasticity, though the results have been largely inconclusive.[60] Racial presumptions, then, continue to mediate our understandings of the skin, as well as our research into difference.

There are as many connotations attached to skin as ways to show, display and mark it. I have not dealt here with the extensive subject of body modification or tattooing, though these are each important ways that the skin has been marked to denote social identity or 'tribe'.[61] Not all forms of communication encoded into the body are deliberate. The emotions we display have become quantifiable and measureable on and through the skin while the study of its diseases has become a major clinical specialism. The skin continues to be regarded as a primary site for emotional displays, in healthy ways like blushing, goose pimples, paling, and tanning, as well as pathological ways through skin conditions related to stress or anxiety. Today there is a specialized term for diseases that involve interaction between the mind and the skin: psychodermatologic disorders ('psycho' referring to the mind and 'dermatologic' to the skin).[62] These conditions fall into three categories: psychophysiologic; primary psychiatric, and secondary psychiatric. Psychophysiologic is the name given to skin disorders that are caused by psychological factors, such as psoriasis and eczema, which were also noted to have emotional causes in Willan and Bateman. Primary psychiatric disorders include self-induced skin problems (such as trichotillomania, which involves pulling out the hair, and delusions that one is covered in parasites). And secondary psychiatric disorders are skin conditions that lead to facial and bodily disfigurement and are understood therefore to have a negative emotional impact. Did Charles Uncle experience emotional distress as a result of his skin disease? The possibility is not considered in his physicians' report, an absence that might reflect changing sensibilities, but more likely reflects its irrelevance to the physicians' immediate concerns.

Whatever interpretations we place on the skin, it grows with us, ages with us, and decays with us.[63] Despite its modern-day status as a psychical and physical boundary that separates the individual from the world, the health of the skin reflects not only our individualism, but also our human need to belong to a broader social world. Without touch, without human intimacy, we fail to thrive.[64] This human need for touch is well expressed in the Japanese and Korean term 'skinship', originally used to describe the intimacy and closeness between a mother and child, but now generally used to refer to bonding through physical contact: holding hands, hugging and bathing.[65] This wider relevance of skinship is a reminder to us of the social nature of bodily experience and the fundamental need for intimacy that is part of the human condition. Despite the skin's role in marking our physical boundaries, we are not unitary selves but social beings, connected to the cultural, spiritual, and material worlds through all our emotions, thoughts, feelings, and senses.[66]

7

Tongue-Tied? From Nagging Wives to a Question of Taste

> That skull had a tongue in it and could sing once. How the knave jowls it to the ground, as if it were Cain's jawbone, that did the first murder! It might be the pate of a politician, which this ass now o'erreaches, one that would circumvent God, might it not? . . . Or of a courtier, which could say, 'Good morrow, sweet lord!' 'How dost thou, good lord?' This might be my Lord Such-a-one that praised my Lord Such-a-one's horse when he meant to beg it, might it not?
>
> (William Shakespeare, *Hamlet*[1])

This scene is one of the most well-known in *Hamlet*. A gravedigger inadvertently and carelessly exhuming other skulls, tossing them to one side, provokes a monologue on mortality from Prince Hamlet, who takes special notice of the skull of Yorick, the late court jester: 'Alas, poor Yorick! I knew him, Horatio: a fellow of infinite jest, of most excellent fancy . . . here hung those lips that I have kissed I know not how oft. Where be your gibes now? Your gambols? Your songs? Your flashes of merriment?'[2] The skull was a common *memento mori* (from the Latin: 'remember that you have to die'), a reminder to early modern men and women to consider the vanity and transience of earthly life.[3] Death is the great leveller; whatever skills and talents one might have in life, whatever powers of influence, they vanish at one's death. This is graphically evident in the absence of skin and flesh that once covered the skull, the missing tongue that once rhymed and jested.

Speech was and is inseparable from the tongue. The word 'tongue' (like the Latin *lingua* and the Greek *glossa*) means 'language' and evokes what one historian has called 'a relation between word and flesh, tenor and vehicle, matter and meaning'.[4] Tongues describe what is said (in the 'mother tongue' of one's native language) as well as the ways in which it is uttered (slippery like an eel, smooth like butter). The act of speaking has long been proof of humans' godly status. Greek philosophers explored the ways in which humankind differed from beasts, citing variously laughter, the ability to distinguish good from evil, the desire for sex in all seasons and walking upright, but by the fourth century they settled upon reason (*logos*) and its associated strengths: most of all speech (*legô*), meaning 'to recount, to tell'.[5] 'The one special advantage we have over animals', recounted the Roman philosopher Cicero, 'is our power to speak with one another, to express our thoughts in words.'[6] During the seventeenth century the French philosopher René Descartes and others pondered whether it would ever be possible to fashion a machine, or train an ape, so that it could speak.[7] These discussions self-consciously drew on wider hierarchies of race and gender as well as literary precedent. Thus in Shakespeare's *The Tempest*, it is the 'savage' Caliban, animal-like and lacking language, who is positioned as a slave and raises questions about colonialism, language, and race in historical literature.[8]

In all traditions, language signifies the operation of the mind. As the German philosopher Ernst Cassirer put it, 'the mind that thinks and the tongue that speaks belong essentially together'.[9] The God-like qualities of being able to think, to imagine, and to articulate are summed up in another of Hamlet's soliloquys: 'What a piece of work is a man, how noble in reason, how infinite in faculties . . . how like an angel in apprehension, how like a god!'[10] However, linking the human tongue with speech as a form of divinity has always been problematic. The tongue is an ambivalent organ, and not always amenable to being controlled. Thus the English cleric Thomas Adams worried about its unruly nature in *The Taming of the*

Tongue (1619), seeing it 'full of deadly poison' and surprisingly strong: 'an arm may be longer, but the tongue is stronger; and a leg hath more flash than it hath, besides bones, which it hath not; yet the tongue still runs quicker and faster: and if the wager lie for holding out, without doubt the tongue shall win it'.[11] Moralizers of all political persuasions, but especially Puritans, preached and pamphleteered against the sins of the tongue that were identified by Adams.[12]

Anatomically the tongue is sexless and without individual agency. Yet its functions are multiple: the tongue was not only an agent of chaos in Puritanical texts, it has also been a shorthand for gender differences; a diagnostic tool in interpreting health and disease; an organ used to express individuality, taste, and aesthetics; a sexual organ, and a gestural one, encoded in signs and symbols across the globe. The oft-ignored tongue has a history rife with social, sexual, psychological, linguistic, and physiological significance. When the 'cat has got your tongue', it is because you cannot or do not speak. If you have the 'gift of the gab' you are 'silver-tongued'—especially if you are a man. Conversely, if you speak through divine intervention rather than volition, some Evangelical Christians believe, you speak 'in tongues', and very little needs to make sense to your fellow believers. According to Corinthians, if you speak in tongues, you speak to God alone.[13] It is not only Christians that value the tongue; the word is also central to the Koran, for example: 'There is a party amongst those who distort the Book with their tongue that you may consider it to be (a part) of the Book, and they say, It is from Allah, while it is not from Allah, and they tell a lie against Allah whilst they know.'[14]

The language of the tongue is part of a richly layered cultural history of gesture. In the UK and the US, for instance, sticking one's tongue out is considered childish or defiant, though it has recently been reworked as a sexualized and provocative gesture in the modern 'selfie'.[15] Yet in Tibet sticking the tongue out has been identified as a traditional form of greeting.[16] Responsible for language, and sensory and gustatory perceptions as well as sensual experience, and endowed with a range of symbolic characteristics, the tongue is as gendered and multi-layered as the rest of the human body. This chapter explores some of its multiple functions through a series of historical episodes that reveal the tongue firstly as a social and political weapon, secondly, as a conveyer of taste (literal and symbolic), and finally as a measure of health and disease from Hamlet's time to our own.

THE TONGUE AS GENDERED WEAPON

Early modern discussions of the tongue, as both Thomas Adams' text and the works of Shakespeare indicate, were primarily concerned with the organ's potential to disrupt the social or political order. Both men and women were capable of the 'sins of the tongue' and treatises like Adams's were intended to teach, to guide, and to instruct the multitude away from slander, blasphemy, and sedition.[17] The prevalence of treatises warning against excessive use of the tongue reflected a commonplace view that the world was full of 'blasphemies, perjuries, flatteries, filthie & abhominable speeches, cursings, raylings and backbitinges'.[18] This should not be so since the anatomical structure of the tongue shows its intended use by God: 'The tongue is small, which should in itself tell us that our speech should be measured.'[19] The tongue conveyed the heart's intent. But the 'minde' was the tongue's guide, and should prevent the tongue's misuse. Moreover, God had put other censorship mechanisms in place in human anatomy. The idea of the 'double barrier of the teeth and the lips' that should guard the tongue's behaviour was an ancient one, dating back to the classical world and prevalent throughout the early modern period.[20] Thus *The Anathomie of Sinne* (1603) reminded readers:

> First he (God) hath made it [the tongue] tender and soft, to signify our words shoulde be of like temper.
>
> Secondly he hath tied it with many threades and stringes to restrain and bridle it.

> Thirdly it is every way blunt, whereby we are admonished that our words ought not to be pricking or hurtfull.
>
> And fourthlye, it is enclosed with a quicke-set and strong rampier of teeth and gummes and with lippes which are as gates to shut it uppe, for feare it should take too much liberty.[21]

These anatomical guidelines were based on theological observation rather than dissection, unlike those of Helkiah Crooke, court physician to King James I.[22] With reference to the work of significant European anatomists Crooke identified the chief function of the tongue in man as speech, and in animals, taste.[23] This is an obvious reference to the possession of reason and language that differentiates humans and animals. Following Galen, Crooke described the tongue as an active organ of taste, 'attracted to sapors as the eye is to colors', sapors meaning the properties of an object, from Middle English Latin, *sapere*, or to taste.[24] Pierre Louis Verdier wrote in *An Abstract of the Human Body* (1753) of the tongue as 'an organ capable of a great number of motions; its substance is almost entirely fleshy, and it has a root and a joint. On its superior surface are several glands, and many nervous papillae which are covered with a very fine membrane'.[25] The tongue was one of the most important of the five senses that 'causes in the soul a particular sensation': in the case of the tongue and palate, that of taste.[26]

I will return to the theme of taste, and the ways in which the tongue has been mapped as a sensory organ below. Firstly, I want to turn to its social uses and abuses during the early modern period, when slander litigation increased exponentially in both ecclesiastical and common law courts.[27] The practice and pursuit of slander suits was gendered. As the historian Laura Gowing has shown in her study of the London Consistory Courts, a high proportion of litigants were women, and they petitioned the courts most often as a response to sexual slander and the imputation that they were 'whores'. A women's sexual conduct was, unlike that of a man, 'at the absolute centre of her integrity'.[28] For instance, the ecclesiastical court records state that Alice Amos 'walked past the house of Richard and Susanna

Symonds. Susanna leant out of the window and called to her, "Thow art a whore. And I sawe my husband stand between thie legs and thow didst put thow hands into his codpiece very rudely." From inside the room Richard himself called "remember the quart of creame, remember the quart of creame!" and looked out to say "I did occupy the myselfe six times for one messe of creame." '[29] Different codes of behaviour therefore shaped male and female experience, and different notions of 'credit' and reputation, with men more frequently accused of inappropriate financial conduct, of being a 'thief' or a 'drunk'.[30]

The use and abuse of the tongue was as gendered as the words that were spoken. In 1706 the Ordinary of Newgate gave his regular account of the *Behaviour, Confessions, and Dying Speeches of the Malefactors that were Executed at Tyburn*.[31] At the bottom of the list were the usual advertisements for books of moral instruction and education. One of those books was *The Management of the Tongue*, with detailed sections that described the temperament and personality types linked to particular verbal stereotypes and listed no fewer than twenty-seven different ways the tongue could be used:

> '1 Of Conversation. 2 The Babbler. 3 The Silent Man. 4 The Witty Man. 5 The Drol. 6 The Jester. 7 The Disputer. 8 The Opiniater. 9 The Heedless and Inconsiderate Man. 10 The Complimenter. 11 The Man who praises others. 12 The Flatterer. 13 The Lyar. 14 The Boaster. 15 The ill Tongue. 16 The Swearer. 17 The Promiser. 18 The Novelist. 19 The Talebearer. 20 The Adviser. 21 The Reprover. 22 The Instructer 23 The Man who trusts others, or is trusted with a Secret. 24 The Tongue of Women. 25 The Language of Love 26 The Complainer. 27 The Comforter'.[32]

As in the case of slander disputes, where the possession of 'good credit' (sexual or financial) was crucial to one's fortunes, a lying and dissembling tongue could wreak havoc on the social order. Evidence of the 'flatterer', the 'lyar', the 'ill tongue', and the 'heedless and inconsiderate man' can be found in court records from the time, both secular and ecclesiastical. The court records of the Old Bailey, the central criminal court of England and Wales, are comparable in detail to ecclesiastical cause papers, for they reveal much about the language, everyday life, and preoccupations of the people involved. A search of the cases brought before the court between 1674 and 1913 shows 1,141 instances in which litigants referred to 'tongues' in ways that suggest the organs took on a life of their own: dissembling, cheating, lying, or charming their victims.[33] In a bigamy case brought before the Old Bailey in 1676, for example, the defendant was charged with having rambled 'up and down most parts of England pretending himself a person of quality, and assuming the names of good families, and that he had a considerable Estate per Annum'. Pretending to be much higher above his station than he was, the defendant wooed and seduced a number of rich widows, 'formally mak[ing] love to them', which he achieved by being 'of handsome . . . presence, and Master of a voluble insinuating tongue'. Once he had persuaded them to marry him 'and for some small time injoy'd their persons, and got possession of their more beloved Estates, he would march off in Triumph with what ready mony and other portable things of value he could get, to another strange place, and there lay a new plot for a second Adventure'. When found guilty, the defendant petitioned for transportation, but instead was sentenced to death by hanging.[34]

This was not the only man to delude a woman with his 'fine tongue', and many prisoners brought before the Courts was condemned by 'vices of the Tongue', including swearing and lying.[35] On 18 April 1694, Francis Dodd was tried for grand larceny for robbing Elizabeth Landsell, a widow, of a haul of gold pieces, including eight pistoles (the French name for Spanish gold coins, sometimes called doubloons), nine twenty-shilling pieces of gold, twenty broad-pieces of gold, and 150 guineas.[36] Landsell claimed that she had lost the gold on Sunday when she was at church, and that it had been taken from a locked drawer in her house. Her maid, however, 'declared upon Oath, that the Prisoner was in the house above two hours, by himself, above Stairs; and when he came down, he told the Maid he had money and Gold enough, and he would carry her to Holland and leave his Wife behind him at London, pretending more Love to her than his Wife'. The maid denied responsibility for any involvement in the crime, telling the courts that the prisoner had 'made Love to her, although he had a Wife of his own, and deluded her with his fine Tongue'. In the end, the prisoner was acquitted for lack of evidence.[37]

While men's tongues were believed to dissimilate and lie, to flatter and to charm, women possessed 'vexatious' or 'sharp tongues', as in the stereotype of the nag Katherina Minola, the eponymous 'shrew' in Shakespeare's *Taming of the Shrew*, who was 'renown'd in Padua for her scolding tongue'.[38] So ubiquitous was concern for women's loose and scolding tongues in early modern patriarchal culture that shaming punishments were devised to control the offender who was labelled as 'shrew' or 'scold'. These included the 'Cucking stool', in which a woman would be strapped and plunged into water to the jeers of the local community, and the 'scold's bridle', a metal contraption that was strapped over a woman's head (see Fig. 9). A bell was sometimes strapped on top to magnify the humiliation. The use of the former is well documented in England, and later Europe, from the 1500s, and it has been suggested that it remained in use until the early 1800s. As a punishment it is seldom recorded in court documentation, however, as never legally sanctioned.[39]

Women's tongues, especially their nagging, were believed to drive early modern men to violence, a theme that is depressingly prevalent in court cases. In 1691 for instance, Thomas Austin, of Christ Church in London, was indicted and tried for poisoning and murdering his wife Katherine by putting white mercury into a pint of Ale. At the time that the crime was committed, Austin was not living with his wife, but arranged to sup with her at the nearest pub, the King's Arms in St Martins. They each ordered a hotpot, which was a traditional winter drink made from brandy, ale, eggs, and sugar. Austin decided that his was not warm enough, so he sent his wife downstairs to the kitchens to get it heated up. While she was gone he

Fig. 9. An early modern German 'scold's bridle' that fitted tightly onto the head and prevented the wearer from speaking.

allegedly placed mercury in her hotpot, which she drank when she came back upstairs. Soon afterwards, she became 'very sick, and vomited, and died within three or four days afterwards'. Before her death she alleged that her husband had poisoned her, though he denied it. When she died, however, it was reported that the surgeon dissected her—in a remarkably early instance of an autopsy—to discover the cause of death and reported:

> the Liver was discoloured, and the Stomach expoliated, and very black in four or five places; which discolouration must be effected by some Corrosive Matter thrown into the Stomach, and must proceed from Poyson. Other Witnesses declared, That the Prisoner was reported to have had more affection for another Woman, than he had for the Deceased Katharine.[40]

At a time when divorce was not permitted by Canon Law, and it was difficult and expensive to secure a legal separation, it was implied that Austin had murdered his wife in order to have a relationship with another woman.

He denied it, and blamed his wife for her behaviour, claiming that his wife 'was a very ill woman, and had an Evil Tongue'. Since nobody had seen him put poison into his wife's drink, Austin was found innocent—despite the autopsy evidence.

Austin's case is not unusual. Women who were seen as nags and used their tongues against their husbands were far less likely to be supported by the community or the legal system, and men were far more likely to be found innocent of their murder. After all, women were allowed to be 'reasonably corrected' by their husband and it was widely believed that a man had the right to chastise his wife physically as well as verbally if she did not behave appropriately in his eyes. The line between what counted legally as legitimate correction and domestic abuse was unclear, and there were many cases where women suffered appalling physical and emotional abuse from their husbands.[41]

In 1718, for instance, William Townsend of the Artillery Ground Spitalfields, was indicted for the murder of his wife Catherine. He was accused of 'beating, kicking and bruising her on the Head, Breast, Back, Belly, Stomach and toes', witnesses providing a catalogue of evidence that the defendant had been seen drunk and violent towards his wife. He ordered her to leave the room when he entered, a room in her own home where she sat, sad and disconsolate. When she refused he punched her and kicked her, swore at her and 'kicked her full on the Face' with such force that a witness 'feared he would Kick out her Brains'. A witness testified that others intervened at this point and told the defendant not to beat his wife, especially since she was believed to be pregnant. Mary Cook, the woman's midwife, later testified that she was later delivered of a dead child. 'Could she but govern her Tongue', it was widely believed by many including the dead woman, 'she might live as happy as any Woman in the World'. Since the defendant had friends and acquaintances that testified to the fact that he had the 'Character of a Peaceable Quiet Man', he was acquitted of his wife's murder.[42]

There is a disturbing parallel between the community-sanctioned punishment of women who 'nagged' and modern-day domestic violence cases, in which men claim their wives' nagging was provocation to violence.[43] The stereotype of the 'nagging wife' is so ubiquitous that it continues to be presumed as justification for violent crime. There is evidence that police officers and others working in the criminal justice system, as well as juries, invoke this stereotype when they fail to support women in domestic violence cases.[44] We retain a flawed and dangerous notion that women's tongues, women's nagging, 'drive' men to acts of violence.

Outside marriage the tongue was no less dangerous. A loose and unguarded tongue was a threat to political as well as social authority—metaphorically and literally the mouth could open and let the enemy in. The historian Jonathan Gill Harris has explored the ways in which the tongue was figured by early modern commentators as a site of disease, opening up and exposing the body politic to external infection. Thus William Averell's *Mervailous Combat of Contrarieties* (1588) warns that:

> Among all the enemies of a common wealth, there is none more pernicious than in the envious tongue of false and lying Papists, who when they cannot by their open practices prevaile to harme or impugne our happie government, go then about by false lying speeches, not alone to slander our state & perswade others to dislike of our government, but also labour by surmising reports and coloured lies, to strike a terror in the hearts of the common people.[45]

Dangerous talk and wicked tongues perhaps reached their zenith in the 1640s, at least in the eyes of the Royalists, as the tide turned against Charles I. The Archbishop of Canterbury William Laud warned that it was blasphemous as well as seditious to speak against the king, asking English subjects to 'let no one whet your tongue or sour your breast against the Lord, and against his anointed'.[46] Other supporters of Charles I lamented the 'strife of tongues', the 'rising up of the tongue', and the 'murmur' of the 'virulent tongue'.[47] Seditious words were particularly problematic in an oral culture.

Because they were reproduced in court, moreover, they reveal the ways in which women also participated in the broadly defined world of political critique. Laura Gowing cites the case of Sarah Walker of Newcastle, who in 1663, just three years after the Restoration, declared she would raise an army to fight Charles II and his hearth tax.[48] In the political as in the social realm, women's tongues were as dangerous as men's, if not more so. Women's tongues were 'a poison, a serpent, fire and thunder', one writer proclaimed, extending the narrative of women as nags to make them something altogether more dangerous. Women also had 'the glibbest tongue', which was a fearful combination.[49]

Women's tongues were not always regarded negatively. There has been considerable historical attention to 'gossip' as a form of female bonding, strength, and community regulation.[50] Scholars of the nineteenth century like Tara MacDonald have shown that even within fiction, 'gossip' operates as a kind of currency or economic exchange between women.[51] And yet throughout history there is a recurrent trope of female tongues being voluble, untrustworthy, a source of gossip, and a potential threat to the social and the political body. This was a theme writ large across the allied landscape of the Second World War, as seen in the propaganda poster campaigns by the British cartoonist Cyril Kenneth Bird, a.k.a. Fougasse. A series of posters entitled 'Careless Talk Costs Lives' discouraged people talking about sensitive material where it could be overheard by spies: in one instance, a man is talking to his wife over dinner while a spy lurks under the table; in another, two women gossip on the bus, unaware that Joseph Goebbels and Adolf Hitler are sitting behind them.[52] In the artist Gerald Lacoste's 'Keep Mum, She's Not So Dumb', a glamorous blonde eavesdrops on the military men around her. Men had to guard their own tongues in the presence of women, especially sexually alluring women, who were clearly incapable of keeping sensitive information to themselves, as gossips or even as spies. Similarly in Lacoste's 'Don't Tell Aunty and Uncle', it is the sexually desirable young woman who is the greatest threat of all.[53]

A QUESTION OF TASTE

The tongue is not only an object of political and social conflict. It is also an agent of taste. As the French gastronome Jean Anthelme Brillat-Savarin explained in his most famous work, *The Physiology of Taste* (1825):

> It is no easy matter to determine the precise nature of the organ of taste. It is more complex than would appear at first sight. Clearly the tongue plays a large part in the mechanism of degustation; for endowed as it is with a certain degree of muscular energy, it serves to crush, revolve, compress, and swallow foodstuffs. In addition, through the numerous tentacles which form its surface, it absorbs the sapid and soluble particles of the substances with which it comes into contact.[54]

Anatomically speaking, Brillat-Savarin added, tongues differ: some people have more 'feelers', than others, which explains how people eating the same food may differ in their enjoyment. *The Physiology of Taste*, subtitled *Meditations on Transcendental Gastronomy* and rich with 'anecdotes of distinguished artists and statesmen' from both Britain and France, marked the gradual encroachment of French gourmand culture on English understandings of 'taste'.[55]

Prior to the eighteenth century, discussion of differences or hierarchies in taste, both as sensory perception and a marker of distinction, were relatively infrequent. Early modern writers did discuss the tastes of different food, comparing the appreciation of different tastes to the digestion of God's word. Thus William Cowper, bishop of Galloway, wrote in *The Anatomie of a Christian Man* (1611) that 'as the mouth tastes the meat, and lets none goe downe to the stomack, unlesse it be approved; so the eare of the godly tastes words, and lets none goe down to the soule which is not from God'.[56] The tongue, like the ear, acted as a gateway to the outside world, letting in what is 'approved' and useful for men and women of all social classes.

In the eighteenth century 'taste' emerged as a form of social distinction, characterized by growing consumerism and commercialization and arguably by a filtering down of standards from the upper to the lower levels of society.[57] Physical or gustatory taste was believed to be further down the scale than the ability to judge the fine arts. With the work of the Irish philosopher Edmund Burke's *Essay on the Sublime and Beautiful* (1757) and the German philosopher Immanuel Kant's *Critique of Judgement* (1790), 'taste' became associated with mental discernment rather than physical appreciation.[58] Compared with the higher cognitive faculties like sight and hearing, the 'lower' physical senses of taste and smell were further from the creator and too animalistic to receive much attention. Bodily taste might have been 'tied to the fleshy organ of the tongue' and its sensory 'tentacles' or 'papillae', but the ability to discern and to judge 'mental taste conceived as a feeling or sentiment of beauty' required a higher level of understanding.[59]

Gustatory taste did, however, give rise to useful metaphors. The English politician, essayist, and founder of *The Spectator*, Joseph Addison, compared critics to tea connoisseurs in his essay 'On Taste' (1711), and later writers took the analogy further: 'The Man of Taste may be considered as a Bon Vivant who is fond of the dishes before him, and distinguishes nicely what is savoury and delicious, or flat and insipid, in the ingredients of each,' wrote 'Mr. Town, Critic and Censor-General' in George Colman's mid-century periodical *The Connoisseur*—but 'at the same time, he may be regarded as the Cook, who from knowing what things will mix well together, and distinguishing by a nice taste when he has arrived at that very happy mixture, is able to compose such exquisite dishes'.[60] The eighteenth-century linking of intellectual discernment and gustatory taste was rooted in the development of nationalism and empire. For the first time in the eighteenth century food supplies were relatively secure, coupled with imperial expansion that brought both new foodstuffs and jingoistic pride to Britain's elite.[61] Ideas about taste and delicacy were bundled up in and produced by concepts of the Enlightenment, as new commodities became available along with new ways of producing and consuming food.[62]

Physical and social taste came together in the sensory capabilities of the tongue. One characteristic of nineteenth-century writing, by contrast to the eighteenth century and its focus on particularity, was the rise of the gourmet or gourmand. These two terms were once quite distinct, referring to a discerning palate on the one hand and greediness on the other, but they were elided by Jean Anthelme Brillat-Savarin, in his celebration of the 'crimson chamber where sits the discriminating judge, the human tongue'.[63] Self-styled Victorian gourmands attempted to base gustatory taste sensation on universal principles. Philosophical discussions of taste, like David Hume's *Of the Standard of Taste* (1757), stressed the importance of subjectivity, and the ability to discern difference. This was of course a gendered movement; while 'urbanization, economic modernization, the widespread circulation of cheaper and more accessible goods—such as clothes—made it increasingly difficult to see men of true quality', a refined delicacy of taste marked out the true gentleman.[64]

It was also in the nineteenth century that the tongue was mapped anatomically and its specific qualities isolated and identified as the cause of gustatory sensation. Before Addison and class-based forms of distinction that were ubiquitous in eighteenth-century culture, taste was explained according to material rather than psycho-perceptual causation, though that tended to be in relation to external qualities—for instance, for the ancient Greek philosopher Democritus, sourness was created by angular shapes, sweetness by spherical shapes, and so on.[65] Aristotle had identified seven basic taste qualities: sweet, sour, salty, bitter (all of which are familiar to us today), as well as harsh, pungent, and astringent.[66] Galen followed these categories, as did other Greek, and Roman and Islamic writers, and scholars well into the eighteenth century, although they took some away and added others; Ibn-Sīnā, for instance, listed five tastes—sweet, salty, sour, bitter, and insipid.[67]

Anatomical research in the eighteenth century identified discrete regions on the tongue responsible for the specific experiences of taste. In his *First Lines of Physiology*, the Swiss physiologist Albrecht von Haller observed how the:

> nature or disposition of the covering with which the papillae are clothed, together with that of the juices, and of the aliments lodged in the stomach, have a considerable share in determining the sense of taste; insomuch, that the same flavor does not equally please or affect the organ in all ages alike, nor in persons of all temperatures.[68]

Haller did not isolate the tongue, but identified the uvula, the palate, and the oesophagus as equally 'affected with taste'.[69] In *A Physical Essay of the Senses* (1750), the French surgeon Claude-Nicolas Le Cat similarly argued that the tongue was not alone in possessing numerous 'papillae' or 'Organs of Taste'. During dissection Le Cat discovered them 'on the Palate, the inner Jaw', and 'at the Root of the Mouth'. Although the tongue might be the *primary* organ of taste, 'the mouth, gullet and stomach' were similarly perceptive, which was why those born without tongues, or those who had lost their tongues through accident, were still able to enjoy food.[70] Le Cat reminded readers that it was the soul that was ultimately responsible for all sensations including taste, however, that 'being a point past all contest'.[71] This position was in direct contrast to Haller, for whom all sensation could be reduced to qualities in the tissues themselves, rather in any external force or spirit.

By the early nineteenth century many of the specific tastes previously associated with the tongue—including 'insipid'—were related to other senses, as physiologists debated what could be mapped onto the tongue. At one extreme the German physiologist Gabriel Gustav Valentin rejected all taste receptors except the extremes of sweet and bitter, though his view was never popular.[72] Despite their differing spiritual emphases, both Haller and Le Cat had identified the precise regions of the mouth where taste could be distinguished in ways that were compatible with later discussions of the 'taste buds', identified in 1867 by the independent work of the German anatomist Gustav Schwalbe and the Swedish surgeon Otto Christian Lovén.[73]

The discovery of taste buds (or 'taste cups' as they were known by Schwalbe; 'taste bulbs' by Lovén) was invaluable to researchers who distinguished particular parts of the tongue as more sensitive than others to specific tastes: in particular the tip of the tongue being associated with sweet tastes, the edges to sourness and the base of the tongue to bitterness.[74] D. P. Hänig, a German PhD student, quantified these general findings in publishing his thesis in the journal *Philosophiche Studien*.[75] Armed with the knowledge that taste sensation seemed to be more densely distributed around the perimeter of the tongue, Hänig explained his investigations into the ways basic tastes were detected differently across the organ. He used sucrose to test for sweetness, quinine to test for bitterness, diluted hydrochloric acid to test for sourness, and salt for saltiness. Although all taste buds detected the tastes, Hänig made a rough graph to suggest that there was some variation in tastes around the perimeter. This publication led to one of the greatest confusions of taste in the twentieth century.

In 1942 the American experimental psychologist Edwin G. Boring discussed Hänig's data in his *Sensation and Perception in the History of Experimental Psychology*.[76] Boring performed calculations that supported Hänig's findings, and transformed Hänig's data to produce a more simplistic map, which in turn was misunderstood and reproduced as showing distinct areas of the tongue dedicated to a specific taste sensation. Henceforth sweetness became associated only with the tip of the tongue, sourness with the sides, and bitterness only at the back (see Fig. 10).[77] In 1974 the scientist Virginia Collins demonstrated definitively not only that taste buds could detect all tastes regardless of their location on the tongue, but also that taste buds could be found in other places, as Haller and others

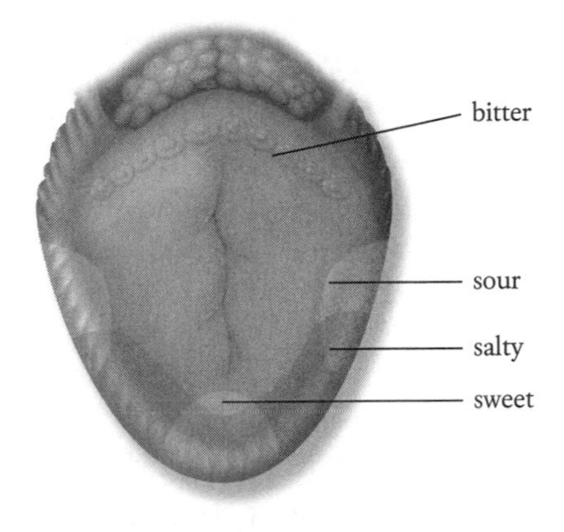

Fig. 10. A now debunked map of the tongue showing localized taste-bud receptors.

had suggested in the eighteenth century. It took some years for Collins' work to become mainstream. Equally neglected was a fifth taste, umami, which was named by the Japanese scientist Kikunae Ikeda in 1901.[78] Umami is found in some Japanese sea vegetables, soy sauce, ripe tomatoes, ripe cheeses, beans, and well-grilled meat as well as the flavour enhancer (and newly recognized obesity creator) monosodium glutamate, which Ikeda isolated and patented.[79]

In the twenty-first century, concepts of taste remain politically laden and culturally specific not only in terms of gastronomic taste, but also in social taste. The ability to discern one taste over another and to exhibit 'good taste' is still the preserve of certain social groups. The French anthropologist and sociologist Pierre Bourdieu has noted this phenomenon in relation to codes and expectations of behaviour, and his judgement of the hierarchies that distinguish one group of people from another.[80] In his discussions of the dietary habits of the French middle class, for instance, Bourdieu noted in his classic text *Distinction* (1979):

> The taste of senior executives defines the popular taste, by negation as the taste for the heavy, the fat and the coarse, by tending towards the light, the refined and the delicate. The disappearance of economic constraints is accompanied by a strengthening of the social censorships which forbid

> coarseness and fatness, in favour of slimness and distinction. The taste for rare, aristocratic foods points to a traditional cuisine, rich in expensive and rare products (fresh vegetables, meat)... [By contrast] teachers, richer in cultural capital than in economic capital, and therefore inclined to ascetic consumption in all areas, pursue originality at the lowest economic cost and go in for exoticism (Italian, Chinese cooking, etc.) and culinary populism (peasant dishes). They are almost consciously opposed to the (new) rich with their rich food.[81]

Gastronomic taste remains hierarchized between rich and poor, educated and uneducated, as well as serving as shorthand for tacitly acquired knowledge. Consider for instance the class associations around wine tasting—like art appreciation, traditionally an aristocratic pursuit.[82] The divide between rich and poor tastes is arguably widening, as food choices and preferences have been linked in sociological literature to a number of economic variables, including income.[83]

A scientific correlation has recently been made between the consumption of salt, fat, and starch, and pleasure sensations in the brain, bordering on addiction, as a result of certain foods. The Pulitzer prize-winning journalist Michael Moss, author of *Sugar, Salt, Fat: How the Food Giants Hooked Us*, explores the ways food giants use terms like 'bliss point' to assess just how much sugar or salt needs to be added to produce cravings and the future desire for more and more of the same.[84] This distortion of the diet in particular in favour of highly processed and calorie-dense products encourages consumers to lose their traditional tastes for nutritious, whole foods in favour of their quick-fix equivalents. This is consistent with findings that our taste buds—ordinarily changeable during the course of a person's life—can be 'dulled' as a result of the amount of processed foods consumed in the West. Links have been identified between heavily processed foods and reduced taste-bud function (and between processed foods and the lower socio-economic groups) that make 'good taste' the preserve of the upper and middling classes. Taste might even be gendered. There is intriguing evidence to suggest that men and women experience food differently according to such factors as women's menstrual cycle.[85]

A BAROMETER OF HEALTH AND DISEASE

> Doctors infer the symptom of sickness not only from a man's appearance but also from his tongue. Surely the most reliable symptoms of a sick or healthy mind are in the tongue.[86]
>
> (Erasmus)

At some stage in our lives, whatever our socio-economic group, we will most likely find ourselves sitting in a doctor's office, being asked to stick out our tongues and say 'Ahhh'. Reading the map of the tongue's health is a well-known and documented way to diagnose a range of health conditions. The use of the tongue in diagnosis was central to Galenic beliefs about health and disease. In humoral medicine the condition of a person's tongue, its colour, wetness, and texture, was used alongside case studies to diagnose the condition of the patient. In *A Bridle for the Tongue* (1663), for instance, ostensibly geared towards curing the sins of excessive or inappropriate language, William Gearing observed how 'physicians take great notice of the tongue, judging thereby of the health or sickness of the body: so our words show plainly the quality of our souls'.[87] The Italian physician Giorgi Baglivi noted in *The Practice of Physick* (1704) that a full medical examination should include 'the Sick Person's Excrements and Urine, his Tongue and his Eyes, his Pulse and his Face, the Affections of his Mind, his former way of living, and the errors he has been guilty of in his way of Conduct'.[88]

Physicians believed that the tongue revealed more than you might imagine about the internal workings of the body. The Danish physician Gerhard van Swieten cautioned that, 'a prudent physician never leaves a patient until he has inspected his tongue and the inside of his mouth, which so fairly shows the state of the viscera... as also of the lungs'.[89] And

Van Swieten's compatriot, the renowned physician Hermann Boerhaave, explained how he focused on the 'eyes, tongues and lips of the patients', using a microscope made from a lens to 'reveal the humors which are visible and not covered by a skin. These lenses are best enclosed in a tube,' he added, 'so that it is not necessary to bring one's face too close to these parts of the patient.'[90] This distancing from the body of the patient was common to the eighteenth and nineteenth centuries, seen also, for instance, in the development of the stethoscope by the French physician René Laennec, who wished to hear the sounds of his patients' chest without getting too close to their bodies.[91] In mid eighteenth-century Edinburgh, too, John Rutherford, Professor of the Practice of Medicine, stressed the importance of examining the face and the mouth of patients, for the gums and tongue could reveal much about their internal humours: 'finding them in a florid state then the blood is in a good state but if they are pale or livid it is a sign that blood is dissolved and watery, if they have a yellowish cast it is frequently attended with a degree of acrimony, but when the cast is of a greenish colour, the acrimony is much greater, as we see in scorbutick people [those affected with scurvy]'.[92]

Throughout the nineteenth century, physicians in Britain and America continued to rely on tongue diagnoses. The casebooks of the physician Peter Mere Latham reveal extensive reliance on medical examinations of the tongue.[93] Physician extraordinary to Queen Victoria and so prolific in the new field of cardiology that he was known as 'Heart Latham', his diagnoses measured the colour, texture, shape, and coatings of the tongue in some detail. Categories included the colour of the tongue ('pink' being the optimum, 'grey' and 'dull' less desirable). The tongue could be 'moist' or 'dry' or 'clean' or 'loaded' (the ideal being 'clean' and 'moist'). Thus Latham described one patient's tongue as unhealthy because 'covered with mucous'. That of another patient, Sir William De Marten, was 'whitened and moist with a slight redness at the edges'.[94] The value of tongue diagnosis in Eastern medicine has traditionally been even more significant than in the West. Indeed, it has been claimed that Western practitioners originally adopted tongue inspection as a direct result of translations of Oriental texts.[95]

It might be argued that by the late nineteenth century, with the decline of the humours, tongue inspection ceased to have any real diagnostic function, though it remained useful as a ritual to display the physician's craft. In other words, tongue diagnosis lost its original status as a barometer of internal health to become part of the gestural repertoire that a patient expected from her physician. In 1871 Sir Thomas Watson warned new medical students at King's College Hospital that 'a patient would think you careless, or ignorant of your craft, if you did not, at any visit, look at his tongue, as well as feel his pulse'.[96] In modern Western medicine, examining the tongue is now used principally in diagnosing localized pathologies like tonsillitis or oral cancer.[97] The examination of the tongue is relatively non-invasive and low-cost and clearly ubiquitous as a social trope for understanding health, even if the socio-medical environment has moved away from illness as a form of bodily imbalance.[98]

Considering the diverse and changing imagery of the tongue, as well as the conventions that surround it—from linguistic conventions and symbols to diagnoses of health and disease—reminds us of the complex and socially situated politics of the body and its parts in history. The tongue is technically not a sexed organ; it does not differ between men and women, except in the attributes with which it is invested. Yet it is layered with gendered presumptions about self-control, social roles, and behavioural codes. So, too, is the sense of taste, which is culturally diverse as well as changing over time.[99]

Questions of taste necessarily lead to appetite and, from the Victorian period, to questions of willpower. Let us turn then to the cultures of the gut and to obesity, one of the central health preoccupations of the modern world.

8

Fat. So? Gut Knowledge and the Meanings of Obesity

This enormously fat man sat in a sofa wide enough for three or four people, and filled it well. He had a really quite handsome, small head, at least compared with his ungainly body. Had he been able to stand up, a feat that really must have been impossible for him to perform, he would have been quite a tall man. His wide cheekbones and huge double chin did not disfigure him very much, but his belly, dressed in a striped waistcoat, resembled a huge featherbed, and his legs, dressed in similarly colored stockings, were the size of two large butter kernels.[1]

THE CASE OF DANIEL LAMBERT

This description of Daniel Lambert was written in 1808 by the Swedish artillery captain Johan Didrik af Wingard, who visited Lambert when he came to Britain to purchase rifles. Lambert was a gaol keeper and animal breeder from Leicester who had just become the heaviest known person in contemporary Britain. At thirty-five years old Lambert had reached the weight of 50 stone (700 pounds) or 318 kilos. Six men of regular build could be buttoned in Lambert's waistcoat, giving rise to his description as a 'human colossus' or 'mammoth'.[2] Lambert became famous for his unusually large size (see Fig. 11). He served four years as an apprentice at an engraving and die-casting works in Birmingham before returning to

Fig. 11. Leicester gaol keeper Daniel Lambert *c.*1806, weighing almost fifty stone and the fattest man then on record.

Leicester and succeeding his father as head gaoler of Leicester Gaol. When the gaol closed in 1805 Lambert, whose bulk meant that he was no longer able even to walk upstairs comfortably, was threatened with poverty. The following year he took up residence in London and placed the following advertisement in *The Times* newspaper:

> EXHIBITION – Mr DANIEL LAMBERT, of Leicester, the greatest Curiosity in the World who, at the age of 36, weighs upwards of FIFTY STONE (14lb. to the stone). Mr Lambert will see Company at his House, No. 53, Piccadilly, opposite St. James's Church, from 12 to 5 o'clock. – Admittance 1s.[3]

Lambert became something of a national treasure during his lifetime. In the British political cartoon 'The English Lamb and the French Tiger' (1804/6), Lambert was pictured tucking gleefully into roast beef and ale, the archetypal diet of the English, while Napoleon dined sparingly on onion soup.[4] Published at the peak of the Napoleonic Wars this kind of satirical representation imagined Lambert as the whole of England, vast and mild with a significant appetite, by contrast to Napoleon as France: thin, mean, and hostile. In another image from the same period, 'Bone and Flesh, or John Bull in Moderate Condition' (1806), a bony Napoleon gazes in awe at the comfortably seated and padded Lambert. 'I contemplate this Wonder of the World,' says Napoleon, 'and regret that all my Conquered Domains cannot match this Man. Pray, Sir, are you not a descendant of the great Joss of China?' Lambert responds that he is 'a true born Englishman, from the County of Leicester', adding that a 'quiet mind, and good Constitution, nourished by the free Air of Great Britain, makes every Englishman thrive'.[5] Thus the cartoon tapped into a series of narratives about British character and politics and about the calm reason of the body of the English man compared to that of the hot-blooded French and Spanish. Perhaps unsurprisingly given Lambert's famed status, the number of visitors he received was 'very great' and their response was overwhelmingly positive:

> To find a man of his uncommon dimensions . . . possessing great information, manners the most affable and pleasing, and a perfect ease and facility in conversation, exceeded our expectations, high as they had been raised. The female spectators were greater in proportion than those of the other sex, and not a few of them have been heard to declare, how much they admired his manly and intelligent countenance.[6]

Charging an admission fee that was out of reach of all but the wealthy, Lambert soon made enough money to return to Leicester where he bred dogs and attended sporting events. Between 1806 and 1809 he made a series of additional fundraising tours before he died suddenly in Stamford in Lincolnshire. According to the *Gentleman's Magazine*, he died in a bed 'of large dimensions' in his ground-floor apartment, weighing more than 52 stone (728 pounds, 330 kilos).[7] The assembly of Lambert's coffin required over ten square metres of wood and was so heavy that it had to be wheeled to the graveside. An obituary in the *Stamford Mercury* imagined the death of his body as one might an automaton that had run out of steam:

> Nature had endured all the trespass she could admit; the poor man's corpulence had consistently increased until . . . the clogged machinery stood still, and this prodigy of mankind was numbered with the dead.[8]

By modern standards Lambert's weight was not so spectacular, even at its peak. One hundred and seventy years after Lambert's death the American Jon Brower Minnoch weighed in at 1,400 pounds or 635 kilograms, almost double Lambert's weight.[9] As this chapter will show, attitudes towards obesity have been transformed between Lambert's time and our own; a change in relation to the status of 'fat' and to the moral loading associated with obesity. For all the contemporary discussion of Daniel Lambert, what is missing is the condemnation, the negativity and even the disgust found in attitudes towards obesity that exist in the modern West.[10] Lambert was 'uncommon', a 'spectacle', and a 'curiosity', rather than a figure of condemnation. A marked shift in framing obesity took place between the early nineteenth and the twentieth centuries, in relation to the definition of 'fat' as a cultural and economic problem, as well as the psychological causes and impact of obesity.

Today we are said to be in the grip of an obesity 'epidemic'.[11] The World Health Organization (WHO) regards obesity as 'one of today's most blatantly visible yet neglected public health problems, with more than one billion people in the developed world categorized as obese: a quadruple growth since the 1980s.[12] Obesity is blamed for a whole range of conditions from heart disease to diabetes, and is identified as the largest killer in the Western world. In 2005 the 'war against obesity' replaced the 'war against tobacco', even though tobacco sales were consistently on the rise.[13] According to the US Department of Health, a third of all adults are obese and a further third overweight.[14] A similar pattern is seen in the UK, with the so-called 'epidemic of obesity' attributed to too little exercise and too much unhealthy food. In 1993, thirteen per cent of men and sixteen per cent of women were identified as obese. In 2011 that figure rose to twenty-four per cent of men and twenty-six per cent of women.[15] The language of an epidemic—and such terms as 'globesity' to describe a global problem—triggers moral and economic panics about the diseases associated with fatness: from diabetes, fatty liver disease, and hypertension to polycystic ovary syndrome.[16] A rise in childhood obesity is perhaps the greatest challenge of all, linked as it is not only with a future of unaffordable healthcare but also the stigma of perceived parental abuse.[17]

Susie Orbach famously declared *Fat is a Feminist Issue* in her classic book of 1978.[18] Fatness is also fraught with concerns about class and ethnicity, and the ways we frame notions of excess around the physical body. It is the mind, too, that comes under scrutiny as the possession of fat is related to psychological issues around hunger, greed, and self-control, the latter of which has come to be identified, more than anything else, as a failing in those that are 'morbidly obese', a classification that did not exist in Lambert's time.[19] Along with anxiety and depression and low self-esteem, the morbidly obese are said to have 'poor impulse control' and appetite regulation, especially those who seek assistance through weight loss or bariatric surgery.[20] This is particularly the case in the calories-in versus calories-out thermodynamic model of obesity that has dominated discussions of obesity since the early twentieth century, in which the calorific value of food eaten is balanced against the energy expended through exercise.

There is a complex, intertwined history of fatness as a social phenomenon, and of the gut as a physical one. Obesity was redefined in the years following Lambert's death, as a social and economic issue as well as a health concern. The categorization of fatness and calories in terms of waste and expenditure, and the association of obesity with greed and idleness (and with the psychological control of bodily appetites) is a subject that is of keen relevance to us today. This chapter draws together cultural ideas about obesity in history with a consideration of the function of the gut. It considers the rise of 'gut feelings' and the changing status of fatness in the West. For in the twenty-first century there is a growing consensus that obesity is not simply about balancing what we eat with how much we move. The redefinition of the gut links to this debate: its nerves, its intestinal flora, and its chemistry potentially reshaping how we feel, as well as our size.

FROM GALEN TO CHEYNE: DIET AND OBESITY

In the mediaeval period, fatness denoted wealth, power, and privilege. In a world of food restrictions, poverty and starvation among the masses, plumpness could be a positive, even a coveted characteristic.[21] Yet even then too much fat was suggestive of gluttony, one of the seven deadly sins.[22] It also carried health risks. The English physician and medical writer Tobias Venner published a handbook of diet and health in 1620 that reminded readers of the importance of humoral balance.[23] A 'fat and grosse habit of the body' was 'worse than a lean for besides that it is more subject to sicknesse', fat bodies 'abound with many crude and superfluous humors'.[24] For Venner, excessive fat obstructed the natural processes of the body and prevented the stability of internal heat that was necessary for optimum health. Rather than being a specific disease in and of itself, in other words, fatness weakened the body and encouraged other diseases. Venner's perspective is reminiscent of that of Hippocrates, who saw 'corpulence . . . not only as a disease itself, but the harbinger of others'.[25] One of the foundational classical

Sanskrit medical texts, the sixth-century *Susruta Samhita* similarly related obesity to diabetes and heart disorders.[26] Like Galen, Susruta, widely regarded as one of the founders of surgery, recommended physical exercise to cure obesity and its side effects.

Galenic doctors saw obesity as a type of inflammation, characterized by changes in the humours and the blood. Importantly, Galen used three different terms to describe what we call obesity: *pachis* (fat), *efsarkos* (εύσαρκος, chubby: from εύ, pronounced'*ef*'in Modern Greek, which means well, and *sarka* meaning flesh), and *polysarkos* (obese, from *poli* meaning much). Galen also identified two different types of obesity: the moderate type that was natural and the immoderate type—polysarkos—that was not.[27] Since Galenic physiology stated that digested food is converted to blood, obesity meant a surplus of bad humours: overeating produced both excess humours and excess fleshiness.

A natural balance was fundamental to humoral medicine. To be too thin was as bad as being too fat, as Venner explained: 'betweene the two habits there is a meane, which is neither too fat nor too leane or extenuated, and that verily is the best because the mediocrity of habit and constitution cannot be but through goodness of the composition which a strong digestive faculty and strength so firme do follow'.[28] Excessive fat or 'exceeding' fat was diagnosed when a man or woman could not walk without sweating or reach the table because his or her belly was in the way, could not breathe easily or clean themselves.[29] Practical obstructions to a person's self-care guided what meant too fat or morbidly obese.

Too much food was to blame. But that still did not mean fat people were ostracized or stigmatized for being obese—any more than they were blamed for other habits of life that gave rise to humoral imbalance. Restrictions on diet, purges and vomiting and bleeding, exercise and massage all helped to combat obesity.[30] In the eighteenth century there was even less emphasis on a person's power of resistance, especially when the possibility of hereditary factors raised their head. In 1757 the Dutch physician Malcolm Flemyng presented an influential paper on obesity to the Royal College of Physicians in London. It was published three years later as *Discourses on the Nature, Causes and Cures of Corpulency*.[31] A 'voracious appetite' was one cause of obesity, wrote Flemyng, but it was far from the only cause, other factors being the 'lax' condition of the body and its failure to excrete waste, especially of animal fats and oils.[32] Moreover, some people seemed to be predisposed towards obesity, especially when it came to the laxness of the body's fibres.[33] Flemyng's work aimed to balance the fluids and solids of the body and its processes of ingestion and excretion. His main recommendation was the consumption of soap to break down body fat. He based much of his physiological understanding on the account of obesity and nervous debility given earlier in the century by the Scottish physician and philosopher George Cheyne (1671–1743).

Perhaps more than any other medical figure Cheyne is associated with the medical treatment of obesity in Georgian Britain.[34] In the first half of the eighteenth century he wrote extensively on self-care and preventative medicine, with such works as his *Essay on Health and Long Life* (1724).[35] Cheyne's patients included the writers Alexander Pope and Samuel Richardson, the Earl of Bath, the Earl and Countess of Huntingdon, and many other well-connected visitors to the spa town of Bath where Cheyne held his practice. In Cheyne's time, the rich were thought to be more sensitive physically and mentally than the poor, and therefore more vulnerable to nervous disorders. Cheyne enjoyed the finer things in life and as his success grew, so did his weight. He weighed about 448 pounds or 203 kilograms at his heaviest and struggled with shortness of breath, ulcerated skin and depression.

In the 1730s, Cheyne lost most of his excess weight by switching to a vegetarian diet. He subsequently published *The Natural Method of Cureing the Diseases of the Body, and the Disorders of the Mind Depending on the Body* based on his experiences.[36] Cheyne also wrote about the problem of nervous debility in *The English Malady*, in which he related a growth in nervous disorders to the wet weather, the damp air, the rich food, urban living, and the inactivity and sedentary lifestyles of the wealthy—who were most likely to suffer from nervous problems like distempers, spleen, the 'vapours', and low spirits.[37] These lifestyle factors, Cheyne insisted, 'brought forth a class and set of distempers with atrocious and frightful symptoms, scarce known to our ancestors, and never rising to such fatal heights, nor affecting such numbers in any other known nation'. Moreover like nerve disorders, obesity was characterized by weak nerves, as:

> A fat, corpulent and phlegmatic constitution is always attended by loose, flabby and relax'd fibres, by their being dissolved and over-soaked in moisture and oil.[38]

The only cure was to tone up and strengthen the fibres with appropriate diet and exercise, by returning the physical structure of the body to its original state. Abstinence, temperance, exercise, and purging (through vomits and emetics) were the means to rebalance the constitution. This is reminiscent of humoralism and Galen's use of the non-naturals: air, food and drink, sleeping and waking, motion and rest, excretions and retentions, and the passions of the soul. Yet Cheyne also drew on theological ideas about controlling the appetite and exercising restraint. In the preface to *The English Malady* he acknowledged that people ridiculed him for the extremeness of his views and the fact that his own weight yo-yoed: that he had 'advis'd people to turn monks, to run into deserts and to survive on roots, herbs and wild fruits . . . that I was at Bottom a mere Leveller, and for destroying Order, Rank and Property' because of his criticism of the lifestyles of the rich. 'Others swore that I had eaten my Book, recanted my Doctrine and System (as they were pleased to term it) and was returned again to the Devil, the World, and the Flesh. This Joke I have also stood.'[39]

Cheyne's religiosity was relevant because he made a spiritual relationship between asceticism, obesity, and Christian behaviour. He saw disciplined food consumption as a religious duty and blamed gluttony on the expanded trade that brought exotic food, drink, and plenty to Britain. Cheyne's brand of asceticism appealed to the patients who lived or attended Bath, with its presumption of a fashionable sensitivity that was in keeping with the aspirational bodies of the elite.[40] His regulation of weight through monitoring diet and exercise was characteristic of approaches that viewed the body as an input and output system, a metaphor that would be even more popular during the industrial age. The Greek term *diaita* originally meant 'way of life' or 'way of being' rather than to indicate restriction on food. But with the popularity of work by physicians like Cheyne, 'diet' became a topic of conversation and commentary among the educated classes: worried over, recorded, and inventoried to the smallest detail.[41]

MEASURING OBESITY, CREATING WASTE

The general timbre of attitudes towards obesity before the eighteenth century was about balance and harmony, regulating the body according to traditional ideals of balance and the non-naturals. Yet by the mid nineteenth century fatness was becoming stigmatized. What was once a matter of Christian ascetics and the resistance of temptation became an economic obligation in the industrial age, especially as it became easier to get fat. In 1650 only the nobility and the wealthy ate sugar. By 1800 it was a necessity in the diet of English people of all social ranks. And by 1900 it was supplying 'nearly one-fifth of the calories in the English diet'.[42]

It was not just diet that was a problem, but the mechanical body's failure to use what it consumed. This was, after all, the factory age and body metaphors were concerned with production, performance, and waste.[43] In the industrial age more than any other, the body was charged up, run down, used up or inefficient.[44] It needed fuel to function, and that fuel was food. Too little food meant the machine would fail; too much food meant waste. The obese body visibly displayed how much energy was being wasted. From the late nineteenth century, that waste could be articulated through the 'calorie'. The term was not formally recognized as a unit of physics until the late nineteenth century when it was widely used following the work of researchers like the German physician Julius Robert von Mayer (1814–78), one of the founders of thermodynamics, a branch of physics concerned with heat and temperature, and their relation to energy and work.[45] Though Cheyne and his predecessors had advised on the relationship between food intake and the structure of the physical body, the use of the calorie made it possible to be far more specific; to make hard and fast correlations between how much food one should eat, and how much one should move.

For the first time too, a number was placed on the exact degree to which one was judged to be overweight or underweight as ideal weight norms were created. Between the 1830s and 1850s the Belgian polymath Adolphe Quetelet developed a practical index to determine an individual's relative size based on measuring a person's weight in kilograms divided by the square of her height in metres. This system was initially termed the Quetelet Index; after 1972 the more self-explanatory term 'Body Mass Index' (BMI) became more popular, as used by the American food scientist Ancel Keys.[46] Though waist measurement is today becoming a more popular measure of health, the BMI is the most prevalently used tool to assess obesity in the modern West.[47]

Insurance brokers arguably did more to popularize the BMI Index than scientists and physicians or governments and politicians. The link between obesity and cardiovascular disease was not newly recognized in the twentieth century; it had been discussed in eighteenth- and nineteenth-century autopsy reports, though 'flabbiness' of the tissues and the heart was often blamed.[48] In the United States, Louis I. Dublin, statistician and vice president of the Metropolitan Life Insurance Company, famously used the BMI tables to classify morbid obesity, and the limitations, therefore, in terms of their insurance: morbid obesity was classified at over seventy per cent the desirable weight of any frame.[49]

The development of nutrition science in Britain and North America was an important factor in understanding obesity as linked to food consumption. This field created by physiologists, biochemists, and physicians, narrowed the dimensions of 'diet' at the same time as it made it more narrowly about the regulation of appetites for the good of the state. The historian Geoffrey Cannon has related the growth of nutrition science to a utilitarian movement by European and American governments to increase food production and build up human resources for war and trade.[50] The malnourishment of the poor was originally as much a focus of attention as obesity; calories were initially used to make sure the general populace got enough food, as well as to prevent them from overeating.

At the same time as norms of weight were established and obesity was regarded as excess, fatness was pathologized as a medical complaint or 'disease' that produced other conditions, as observed by William Harvey, Surgeon to the Royal Dispensary for Diseases of the Ear in Soho, London. After learning of the French physiologist Claude Bernard's experimentations with glucose and diabetes, Harvey started to recommend low-carbohydrate principles to his patients, anticipating the Atkins diet of the twentieth century.[51] One of those patients was the funeral director William Banting (1796–1878). Banting stood at five feet five inches tall and weighed more than 202 pounds or 91 kilograms. Banting was one of the most eminent funeral directors in Britain, and yet he could not bend to tie his own shoes without extreme difficulty and was 'compelled to go down stairs slowly backwards, to save the jar of increased weight upon the ankle and knee joints'. Before he worked with Harvey, Banting was obese and miserable. 'Of all the parasites that affect humanity,' Banting lamented, 'I do not know of, nor can I imagine any more distressing than that of obesity.'[52]

In 1864 Banting read with interest an article on obesity in the Victorian literary journal *The Cornhill Magazine*. The article entitled 'Corpulence', written by the English doctor and writer Francis Edmund Anstie, addressed 'all classes' on the social problem of obesity.[53] The 'celebrated Daniel

Lambert' was exceptional, Anstie argued, as 'bodily and mental activity' was often reduced as a result of fat.[54] Although it was one of the most 'useful tissues of the body', performing the necessary function of a 'cushion, filling up the spaces between more important organs', too much fat could be injurious to health.[55] Most of the blame for obesity, he continued, could be attributed to the consumption of fat, as 'fatty food might generate fatty tissue'.[56]

Banting was frustrated by this article, and by physicians' and the general public's apparent lack of understanding about obesity. After years of enduring the jibes and snide comments of other people, Banting met with Harvey, who placed him on a diet of meat, fruit, and alcohol. In just over a year he lost 46 pounds in weight (nearly 21 kilograms) and over 12 inches from his girth, as detailed in his pamphlet *Letter on Corpulence* (1864).[57] Moreover his general health, including his sight and hearing, the problem that had initially taken him to Harvey, had improved immeasurably. Banting concluded his work with a copy of a BMI table, given to him by an insurance company and used to predict the health and life expectancy of those insured.[58]

Banting's pamphlet was so popular that it brought into everyday language the phrase: 'do you Bant?'[59] As with most other diet books, Victorian or modern, the emphasis was on the removal of certain foods or food groups (especially potatoes, bread, milk, and beer). It did not suggest that obesity was about a lack of willpower or self-control, but about a lack of education and understanding. Yet the importance of self-control was far more commonly stressed by those concerned about the obese, as willpower became a moral equivalent to industrialized power, requiring force and energy and drive.[60] In Britain Samuel Smiles, author of *Self-Help with Illustrations of Character and Conduct* (1859), which was popular on both sides of the Atlantic, noted the importance of 'perseverance, application, and energy...along with persistency and practice'.[61] He reminded readers that patience was everything, after all, the world's greatest achievers, from Isaac Newton to the Comte de Buffon, were untiringly self-denying. Self-denial and self-control were transformed into spiritual and economic virtues.[62] This emphasis was characteristic of the individualism and self-reliance expected in a capitalist consumer economy. The implicit attribution of psychological weakness to obesity is reminiscent of the feminist philosopher Susan Bordo's work on anorexia.[63] Over-eating, like under-eating, is part of a bigger story about individual containment and the limits on the body's physical boundaries. Obesity, like thinness, signals a perceived failure in self-regulation and a deviation from a socially approved and standardized norm. This is especially problematic for women, whose bodies are more commonly exposed to comment and censure than their male counterparts.

GREEDY GUTS? OBESITY AS A MORAL OFFENCE

Today there is no status in being fat in a time of plenty, except in countries like Mauritania, where resources remain scarce and wealthy individuals deliberately overeat in order to achieve a pleasing degree of obesity.[64] In the UK, about forty per cent of women and twenty per cent of men are said to be dieting at any one time.[65] The diet industry is booming, from the Atkins to the Zone, from WeightWatchers to the South Beach Diet. And yet ninety-five per cent of all diets fail.[66] Given that the food giants that produce highly calorific, intensely processed foods with chemicals to maximize 'mouth feel' and 'repeat appeal' (otherwise known as addictiveness) are the same organizations that are behind the slimming industry, willpower alone is clearly not enough.[67]

Adverts for weight-loss remedies were widespread from the nineteenth century in the United States and Britain as new pockets of leisure and disposable income were created.[68] Products like 'Allan's Anti-fat Remedy' and periodicals aimed at the white professional classes boasted cartoons that depicted fat people as greedy and self-serving creatures worthy of ridicule.[69]

Fatness was stigmatized just as the cult of dieting and thinness was established, with leanness becoming the cultural ideal. Obesity became a social marker: poor people subsisting on quickly produced, cheap, processed foods were fatter than their wealthy counterparts with the time, interest, and the education to prepare 'slow food'; a conscious, middle-class response to the fast-food revolution.[70] Today it is the fat poor who are mocked and blamed for being overweight and targeted with health campaigns focusing on willpower and exercise. It is widely recognized that fat people are not only more likely to suffer heart disease, diabetes, and cancers, but also to take medication for those diseases. According to the *Irish Medical News*, it is time to take a deliberate decision to 'stigmatize' obese people, in much the same way as we ostracize smokers.[71]

BEYOND THE THERMODYNAMIC MODEL

Fatness cannot be regarded as a simple measure of calories-in versus energy-out, and it is outmoded to ascribe obesity to a lack of willpower. When we feed our bodies we are not simply fuelling an automaton; our bodies are complex biochemical and psychosocial entities. Today the mechanical metaphor feels redundant, and yet we still describe bodies as machines. The hydraulic metaphors from a much earlier age are used to describe not only the process of eating (we run on empty and need to refuel), but also our emotional states; how often are we 'pent up' or 'ready to explode'?[72] Perhaps we are too close to our own period to be able to imagine the body metaphors that will be appropriate for the digital age and beyond.

In the twenty-first century, attempts to tackle obesity as a governmental and social problem tend to focus still on psychological control, whether that is curbing the appetite through willpower and appetite suppressants, providing guidance on diet and exercise so that people make healthier choices, or—more controversially—offering a range of financial and other incentives.[73] And yet 'emotional eating' is widely recognized as a problem, especially among women, along with 'mindless' eating and the over-consumption of processed foods.[74] The correlation between psychological need and hunger is one that has not been adequately tackled. Also neglected is the role of the body in generating the need for food above and beyond what it needs to survive. The thermodynamic principles behind the calorie model are therefore challenged by biochemical interpretations that focus on the body's chemical processes, proteins, and hormones as responsible for obesity.[75]

There are perhaps two separate but related issues at work here. The first is what Bordo has described in another context as the 'disordered consumers' experience.[76] The economic, social, and ideological conditions that foster obesity are multiple and there are arguably vested interests in keeping it that way. These include 'big business': food manufacturers, advertisers, fast-food outlets and so on, as well as a pharmacological industry that essentially profits from the obesity 'crisis'.[77] Moreover, the separation of mind and body in Western medicine is as unhelpful in tackling obesity as the poor nutritional content of many diets. This is no longer just a Western problem; a 2012 article in *Nature* argued that every country that has adopted the Western diet dominated by low-cost and highly processed food has seen rising obesity rates and associated diseases.[78] It is sugar that is seen to be the problem rather than the archetypal enemy, 'fat'.[79] Excessive sugar consumption can induce insulin resistance, which basically prevents the body from recognizing we have eaten enough, as well as metabolic syndrome, a cluster of biochemical and physiological abnormalities associated with obesity such as diabetes and high blood pressure or hypertension.[80]

A further issue to consider is the role of psychological factors in weight gain. Biochemical narratives increasingly see fat as a psychological safety-blanket, an emotional defence against post-traumatic stress disorder, depression and anxiety, and a lack of security in everyday life.[81] The bigger we are, in other words, the safer we feel. The idea of emotional obesity is not a new one, but it is normally understood as a psychological desire to eat, rather than a physiological desire to stay fat. A 2006 article in the journal *Molecular Psychiatry* indulged in a little 'ghost writing', imagining what Sigmund Freud's response would be to the globesity epidemic. The principles of psychoanalysis, with its emphasis on early traumas and the fear and anxieties leading to stress and comfort eating were highlighted by the authors, along with a 'paralysing lack of physical activity and eating that is dissociated from actual calorie needs'. In this view it is the combination of readily available 'comfort foods' with poor standards of nutrition, the mass commercialization of desire and 'repressed fears and unconscious conflicts' that produce the obesity epidemic.[82] Moreover, stress and anxiety, like junk food, increase the amount of cortisol in our systems, a steroid hormone that regulates our stress response. Too much, it has been argued, can make us fat.[83]

The drive towards seeing the soma as responsible for holding on to fat turns traditional arguments about willpower on their head. It also relies on particular narratives about the biochemical body that further elevates the importance of hormones as the modern-day humours, responsible for all aspects of our mental and physical health. There is considerable resonance between the idea that our bodies might hold on to fat for psychological reasons and the claim by cellular memory theorists that thoughts and feelings physically impact on our glandular, hormonal, and nervous pathways. There is further resonance with scientific attempts to rethink the body's digestive processes, especially when it comes to the gut. Once viewed mechanistically as a space where food is digested and processed, the gut has been revised as an intelligent, emotional space that provides biochemical feedback to the brain and to the rest of the body. In other words, there are more to 'gut feelings' than meets the eye. We are moving towards a new understanding of the body in which the state of the gut might even hold the key to obesity itself.

OBESITY, DIGESTION, AND THE EMERGENCE OF 'GUT FEELINGS'

Thus far I have focused on the belly principally as a site of consumption. And yet the gut is also linked to our languages and experiences of emotion. Intuitive forms of knowledge—'gut feelings'—are part of a much longer history of experiencing extra-sensory knowledge through the body rather than the mind. The word 'intuition' (from the Latin *intueri*, meaning to 'consider' or 'look on') was first used in the mediaeval period to mean knowledge derived from spiritual insight, or knowledge beyond the senses, and it has been part of the philosophical framework of thinkers from Socrates and Pythagoras to Baruch Spinoza and Immanuel Kant.[84] For Galen, intuitive knowledge was arrived at through the operation of the soul.[85] In Eastern culture too, intuition has been important. It was at an earlier stage than in the West, moreover, that intuition became linked to the gut. In Japanese culture, for instance, the belly is the seat of wisdom and the centre of gravity, both physical and spiritual. The 'hara', the place of balance just above the navel, is also the place of higher thought, the soul and the spirit, and the organ linked to emotions and personality. This difference is indicated by cultural references that centre on the gut, comparable to the ways the heart has resonance in emotion rhetoric: 'hara ga tatsu' ('my gut stands', meaning 'I am angry'); 'hara ga dekita hito' ('a person that has a developed gut', meaning a person with strong moral fibre).[86]

Of course there is plenty of mysticism in North American philosophy. But there is also an ambivalent attitude towards the gut in Western medicine. We accept that emotions and the gut are necessarily linked: we lose our appetites when distressed, anxiety can be linked to constipation or diarrhoea, and love conjures up 'butterflies' in our bellies. But the gut has more often been associated with animalism and brutality than the spiritual and the divine. The process by which emotions and the gut were linked, and the gut became associated as an active rather than a passive recipient of emotional change is one we need to examine more closely. It is not only attitudes towards fat people that has changed since Lambert's time, but also beliefs about the gut and its digestive processes.

This change is reflected firstly in the metaphorical language about the gut and secondly in medico-scientific understandings of how it functions. According to the *Oxford English Dictionary* the first use of the term guts to mean the visceral contents of the abdomen, specifically bowels, was recorded about 1000 CE.[87] 'Guts' shifted—sometimes the term denoted the stomach and other times the intestines. Their metaphorical use mutated. By the early seventeenth century the term 'guts' denoted labour and strength (working your guts out) as well as a way of expressing antipathy towards a person: 'Ile make garters of thy guttes, Thou villaine' (Robert Greene, *James the Fourth* (1598), III. ii). A proverb popular in the sixteenth century was that to have 'guts in his brains' was a valuable asset.[88] There are earlier records of the stomach in particular being a site of courage. In Shakespeare's *Henry V* for instance, the eponymous monarch declares: 'He which hath no stomach to this fight, Let him depart.'[89] And the strength of the stomach was gendered, as seen in Elizabeth I's famous declaration that though she had the mind and body of a mere woman, she had the 'heart and stomach of a king'.[90] In significant ways the stomach and the digestive processes have been gendered in history, the organs of digestion being layered with stereotypes of gender, ethnicity, and class in much the same way as the brain.[91] Complex associations were also made between mental and digestive processes. In the nineteenth century putting your 'guts into it' meant working hard. By the following century, when nervous and chemical connections were made between the brain and the stomach, particularly in terms of stress, gut feelings became explicitly linked to intuitive knowledge: thus in 1969 *The Times* could call 'the moon programme . . . a gut issue'.[92]

Changes in understanding the stomach and digestion enabled this linguistic transformation of the 'gut'. According to Galen, digestion involved an active agent, 'a stomach that is in accord with nature', that 'attracts to itself from above through the cardiac orifice and expels downward'.[93] The stomach, like the heart, had a life of its own that came from the spirits: it could feel its own emptiness and generate the sensation of hunger. From studying the muscles inside animals' stomachs, Galen believed that the stomach squeezed the food. He used metaphors of agricultural labour to envisage 'this storehouse' of the stomach as a 'work of the divine, not human art' which:

> Receives all the nutriment and subjects the food to its first elaboration, without which it would be useless and of no benefit whatever to the animal. For just as workmen skilled in preparing wheat cleanse it of any earth, stones, or foreign seeds mixed with it that would be harmful to the body, so the faculty of the stomach thrust downward anything of that sort, but makes the rest of the material, that is naturally good, still better and distributes it to the veins extending to the stomach and intestines.[94]

Galen believed that food was converted into blood. He presumed that digestion started in the stomach, where food was minced and processed and essentially cooked by its internal heat. This automatically meant, since women were cooler than men, that men's stomachs and digestive processes were more efficient. And since the stomach was also involved in the production of humours (and in a person's emotional and psychological temperament), hotter stomachs converted food into 'more perfect' male bodies.[95] From the stomach food went to the bowel, where it was decomposed before being sucked through the veins and transported to the liver. In the liver food was changed into blood, which then moved to the heart. The amount of blood that was in the body was directly proportionate to the amount of food one ate, a belief that remained intact until the Renaissance.[96]

Early accounts associated digestion problems with stomach diseases, or 'bad' fluids, often as a result of an imbalance in the humours that rose from the stomach. This could cause a range of physical and emotional symptoms. In *The Anatomy of Melancholy* (1621) for instance, the English scholar and divine Robert Burton discussed the 'black fumes' that rose to the brain and clouded the senses.[97] The stomach was the 'king of the belly' for Burton, because when the stomach was out of sorts the rest of the body suffered. Particular types of food, like hare meat, were more likely to cause melancholy than others, because they created thick, black humours. This was the same thinking that inspired Cheyne's pursuit of a vegetarian diet.

These ways of thinking about the stomach are alien to modern medical sensibilities. Today the stomach is seen as a muscular hollow organ that takes in food, mixes it, breaks it down and passes it to the small intestine. The entire digestive system starts at the mouth and finishes at the anus–it is one muscular tube of which the stomach is just a part.[98] The stomach wall is composed of mucous membranes, connective tissues, blood vessels, nerves, and muscles that move the contents of the stomach around and help grind solid food into usable pulp. Digestive enzymes, hydrochloric acid, mucus, and bicarbonate line the stomach to break down its food and aid in digestion. Rather than food being crushed and squeezed in the stomach, it is broken down by chemical processes, an idea that first emerged in the eighteenth century with good and bad ferments–as in the work of the Flemish physician and chemist Joan Baptista van Helmont.[99] Van Helmont was particularly interested in the processes of digestion, and he analysed five separate stages through which it took place. Though his work was problematic it marked an important stage in understanding digestive processes, and moved away from Galen's concept of the stomach 'cooking' food.[100]

Studying the operation of stomach acids in a living human being was the next step to understanding digestion in its modern gastroenterological sense. In the late eighteenth century the Italian naturalist and physiologist Lazzaro Spallanzani obtained the stomach juice from living participants using a sponge on a thread, which they swallowed and kept in their stomachs for some time before the thread was pulled up and the sponge removed for examination.[101] But it was the work of the American army surgeon William Beaumont that took gastric experimentation to a new level.

In *A Modern History of the Stomach* (2011) the historian Ian Miller recounts Beaumont's treatment of Alexis St Martin, a French-Canadian working as a voyager for a fur-trading company in Michigan (see Fig. 12).[102] In 1822 St Martin was accidentally shot in the torso. The wound was serious and perforated his stomach. Beaumont attended St Martin and expressed surprise that the man had survived. He supervised St Martin's care and noted how a year after St Martin's injury his wound had healed almost entirely, aside from a 2.4 inch (or 6 cm) hole in his torso. Beaumont discovered that he could examine the workings of St Martin's stomach, sticking his finger in and introducing muslin bags that he was able to retrieve at will. This allowed Beaumont to see the digestive processes at work. Perhaps unsurprisingly given St Martin's experimental value, Beaumont continued to work with him when charitable support ran out, even taking St Martin into his household as his servant.[103] Beaumont's writings on digestion in terms of chemical decomposition were widely popularized. They showed not only that digestive processes were a matter of chemical breakdown with stomach acids, but also that the emotions and the gut were directly connected. When St Martin had a fit of 'violent anger'—which we can only presume was not connected to Beaumont's relentless fiddling with his stomach—Beaumont observed how yellow bile appeared in his gastric juice, showing 'the effect of violent passion on the digestive apparatus'.[104]

By the nineteenth century the workings of the digestion was a British obsession, giving rise to metaphors and images that related a functioning stomach to a stable person, as well as a stable nation.[105] These nationalistic concerns intensified as the industrial age gathered momentum and the body was supposed to run like an ordered machine—a factor that, as discussed above, influenced attitudes towards obesity. Conversations about

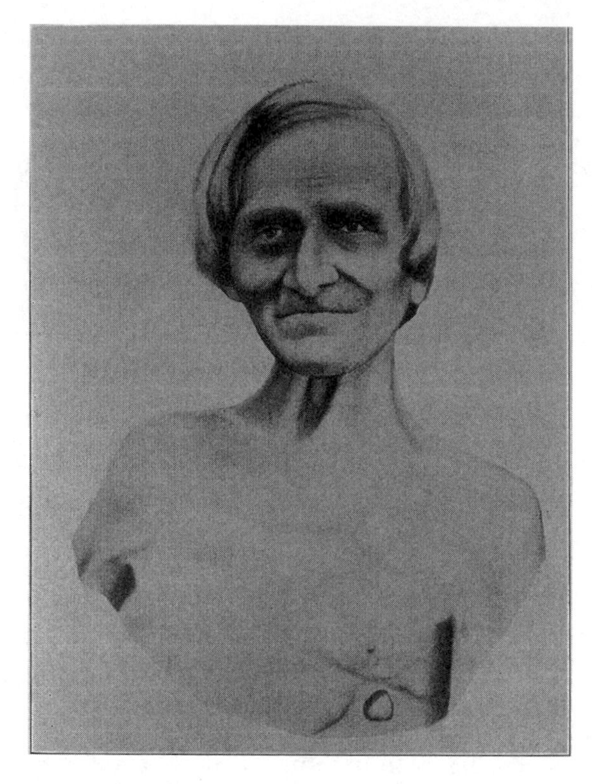

Fig. 12. Portrait of Alexis St Martin, showing the location of the fistula that led to his stomach.

gut problems became as much a part of the national diet as roast beef. 'Stomach diseases are of every day occurrence; they form the national malady of Britain, and consequently the prime staple of the medical art', observed a writer in the *Dublin Medical Journal* in 1838.[106] Against increasing numbers and types of diseases associated with the pace of industrialization, urbanization, and overly rich food, gut problems were especially the province of the élite. The more sensitive and well bred the individual, the more prone he (and it usually was a 'he') was prone to gut problems.

By the early twentieth century high anxiety states were associated with stomach problems, and 'ulcerative' personality types were identified:

hard-working, tense, wound-up individuals who found it hard to relax. The American physiologist Walter Cannon argued that 'just as feelings of comfort and peace of mind are fundamental to normal digestion, so discomfort and mental discord may be fundamental to disturbed digestion'.[107] Freudian psychoanalysis confirmed these concerns by identifying specific psychopathologies with constipation and excretion.[108] While anxiety had been linked to physical illness in earlier historical periods, this explicit association with the gut was relatively new. It was the heart, rather than the stomach that had been traditionally linked to emotional anxiety and excessive work.[109] The association with anxiety and the gut came about at the same time as the language of 'gut feelings'. So did the term 'gut check'—meant as a way to 'assess one's feelings regarding a course of action, typically intended to reconfirm one's enthusiasm or resolve'.[110]

TOWARDS THE BRAIN IN THE GUT

One of the dominant narratives of this book is the secularization of the body; the removal of the sacred (the soul), and the substitution of material over spiritual explanations for how it worked. Another is the way in that the brain qua mind has taken the place of that soul, so that intuition and gut feelings are seen as neuroscientific experiences. There is a mass readership for books like *Blink: The Power of Thinking Without Thinking*—though there has also been a 'slow thinking' backlash.[111] Neuroscientific accounts tend to explain gut feelings with reference to brain-activated responses. The American neuropsychiatrist Louann Brizendine, for instance, author of a controversial book about male and female brain difference, suggests that intuitive feelings spring from a part of the brain called the insula, the neural centre for empathy and self-awareness. The insula is connected to the vagus nerve, which runs through the abdomen and regulates the functioning of various organs, including the stomach and intestines.[112]

Brizendine's approach takes for granted that the brain and the body are linked not only in gastrointestinal function (as was established in the nineteenth century) but also in intuitive decision-making. There is constant 'crosstalk' between the brain and the gut that monitors digestive processes as well as the perception of emotions. More recently this crosstalk has been interpreted as gut- rather than brain-led. In the same way that the identification of a 'little brain' in the heart has led to claims that the heart might actually drive rather than respond to emotional states, gut feelings have been given a material basis.[113]

The American physiologist Michael D. Gershon argued in the 1990s that the Enteric Nervous System (ENS), one of the main divisions of the nervous system that governs the gastrointestinal system, was effectively a 'second brain', based on its size, complexity, and similarity in neurotransmitters and signalling molecules between the gut and the material brain.[114] The ENS contains between 200 and 600 million neurons, the same number of neurons that have been identified in the spinal cord. Moreover, the intestinal surface area is the largest body surface and is approximately 100 times larger than the surface area of the skin. Like the skin, it is claimed, the gut has a considerable role to play in receiving information and in communicating to the rest of the body. In this approach the gut is reframed as a communication centre, signalling to the brain through mechanical pressure, hormones, and the immune system, with the gut containing more than two-thirds of the body's immune cells.[115]

Yet the identification of the ENS, and a non-cranial brain, is not new. The American physician Byron Robinson identified the ENS in *The Abdominal and Pelvic Brain* (1907).[116] Robinson argued that the abdominal viscera contained a vast, complex nervous network that influenced and regulated many of the body's systems, and that mammals possessed both a 'cranial brain' that was 'the instrument of volitions, of mental progress and physical protection' *and* an 'abdominal brain, the instrument of vascular and visceral function':

> It is the automatic, vegetative, the subconscious brain of physical existence. In the cranial brain resides the consciousness of right and wrong. Here is the seat of all progress, mental and moral... However, in the abdomen there exists a brain of wonderful power maintaining eternal, restless vigilance over its viscera. It presides over organic life. It dominates the rhythmical function of viscera... It has the power of a brain. It is a reflex center in health and disease... The abdominal brain is not a mere agent of the [cerebral] brain and cord; it receives and generates nerve forces itself; it presides over nutrition. It is the center of life itself.[117]

At about the same time as Robinson was arguing for the importance of the 'abdominal brain', the British physiologist John Newport Langley was working on the ganglia of the gut. He observed that the nervous system of the gut was capable of functions independent of the central nervous system (CNS).[118] The work of Robinson and Langley was overlooked for several decades before the recent declaration of the second brain in the gut. And until recently the ENS had been seen as subsidiary to the central nervous system, though that hierarchy is shifting.

These findings give scientific credibility to well-established relationships between gastrointestinal disorders like irritable bowel syndrome (IBS) and psychiatric disorders of all kinds. More contentiously the flow works both ways: the gastrointestinal system impacting on the chemistry of the brain in much the same way as the brain impacts on the functions of the gut. Rather than experiences like stress being transmitted from the brain to the belly, which is the conventional top-down approach to conditions like IBS, the physical condition of the stomach and guts and their nutritive and hormonal makeup is seen to influence psychological and psychiatric conditions. And hormones and chemicals previously thought to exist only in the brain have been identified in the gut.[119] About ninety-five per cent of the body's serotonin, for instance, a neurotransmitter linked to appetite, sexual behaviour, pain, and happiness, is found in the gut. So is dopamine, a hormone associated with pleasure and the reward system. In our hormone-dominated age, these fluids matter to scientific discussions about everything, from sexual development and identity, metabolism, growth and development to the body's response to stress. Since neurotransmitters from the gut can influence the brain in happy as well as unhappy states, stimulation of the vagus nerve has been mooted as a treatment for depression.[120] The existence of these gut–brain signals can also help explain why fatty foods are so addictive and why they make us feel good. Fatty acids are detected by cell receptors in the gut lining that send positive nerve signals to the brain.[121] By contrast, too few fatty acids link low-fat diets and depression.[122]

From this perspective our inability to 'stomach' a situation, or the 'knots in our stomach' that tell us something is wrong, can be linked to nervous sensations rising from the ENS or the gut–brain as much as from cognition and the mind. Yet there has been an even more surprising claim. A 2011 study in the journal *Neurogastroenterology and Motility* suggests that much of the work in generating information from the gut is carried out by the intestinal microbiota, or germs.[123] There are approximately 100 trillion bacteria in the intestines, and their presence is often accompanied by neurochemical changes in the brain. Is it possible that depression and anxiety might ultimately be reframed as bacteriological conditions, rather than psychological ones?[124] Hormones in the gut, and bacteria in the gut, have both been linked to obesity.[125] Being fat, then, might not only tell us about how much we eat and how much exercise we get, but also more profoundly about the health of our bodies and our minds.

The development of the 'gut brain' as a site of knowledge, intuition, and feeling, and the relationship of gut flora and hormones with obesity and depression offers a more holistic understanding of the mind–body relationship than is found in conventional Western medicine. The comparison with the trajectory of the heart might be useful, for, while neuroscience seems to answer all the questions we have about selfhood and identity, alternative stories are coming to the fore. Thus far, the heart and the gut provide frameworks for alternative versions of the self. Unlike the language of the heart that originated in the classical period, the language of the guts is a complex blend of modern and ancient models. In the nineteenth century, there was more interest than ever before in the mechanisms of digestion and waste and in the vilification of fatness. Processes of digestion and excretion became linked, with the emergence of the mind sciences, to the operation of the individual psyche. The language of the guts, like that of the heart, has become a way to talk about bodily processes that is instinctual and visceral rather than abstract or philosophical.

What are we to make of these developments, and how do they impact on our views of the self and the mind–body relation? The re-evaluation of obesity as a product of floral imbalance is a challenging proposition. So, too, is the suggestion that psychiatric problems might be linked to what we eat—not least because it is a return to the Galenic principles of humoral medicine and its rules about avoiding certain foodstuffs in the pursuit of optimum health. In the humoral model, melancholic humours could overwhelm the brain; in the biomedical model, foods like sugar impact on our brain chemistry and produce negative physical and psychological states. There is something inherently holistic about reconnecting the mind and the body through the mechanism of fat, and in viewing health as a product of internal fluids—whether they are humoral, chemical, or hormonal. Though our attitudes towards obesity might have shifted since Lambert's time, this kind of holism suggests a socially contextualized awareness of weight as a product of forces other than overeating and a lack of willpower. At a biomedical level this could mean the identification of obesity as a disease to be treated by medications that counter the effects of insulin resistance.[126] More promisingly it could open up the possibility of a more holistic view of the self and a relation between mind and body that is about harmony and balance rather than conflict and control.

Fatness, and its interpretation, speaks volumes about our relationship with food and our bodies, just as eating tells us much about our relationships with others. Changes in viewing the body and the meanings of its parts reveals economic, social, political, and cultural concerns as well as ideologies of class, gender, and ethnicity. Each of the body parts looked at in this book has multiple meanings that have changed over time. The stories told here are not intended to be exhaustive, but merely to serve as illustrations of the complexity of the body and its parts in history. Important themes have been raised along the way—about mind and body, about the limits and the perfectibility of the self, and about the languages used to talk about the body. How we put these disparate parts back together is the subject of the conclusion.

Conclusion

Towards Embodiment

> Matter extends into space. Mind does not. Matter can be divided into pieces, as in a mortuary dissection. Mind cannot . . . mind thinks. Matter does not and cannot.[1]

The French philosopher René Descartes' proposition, summarized above, is that mind is something other than the body—something that thinks and cannot be divided. The relationship between mind and matter has occupied theorists from classical times to the present. The mind has always been associated with the soul, though it did not always encompass it. Sometimes the mind was expressed *as* the soul, reflecting the original meaning of the word psychology (from the Greek *psyche*, meaning the soul or the breath of life). The location of the soul was also uncertain. Early Egyptians placed the soul in the heart. By the seventeenth century the heart was mechanized and the soul moved to the brain. Descartes was specific enough to locate it in the pineal gland of the brain. In the nineteenth century the brain was reduced to material processes and the soul became 'mind'. Yet there is no easy evolutionary narrative that takes us from the heart in the seventeenth century to the material brain in the present. There are many writers who condemn 'Descartes' error', identifying emotional input, for instance, as central to rational decision-making.[2] Moreover, it is not only the heart and the brain that have been associated with intuitive knowledge: the belly has also become a scientifically sanctioned site for 'gut feelings'.[3]

The relationship between mind and body has been one of the main themes of this book, especially in terms of the location of the soul, and the self. We live in an age where the brain is normally conceived as containing the essence of our selves (our minds, that is), but we still regard other parts of our bodies—our hands and our faces, for instance, on the basis of much transplantation debate—as integral to our 'selves'.[4] How we feel about our physicality moreover, and how those feelings are manifested in medical, literary, religious, and other discourses, can be complex and contradictory. I started this book because I was interested in looking at the body as an assemblage of parts, and why it was that some parts take on particular significance and meanings at certain points in history. *This Mortal Coil* has explored some of the multiple meanings given to body parts in the past and in the present. Some body parts, like the tongue, intrigued me because at first glance they seemed to have escaped the same layering of gender and social and political meaning that shroud other organs, though of course they have not. Others, like the female breast, I considered because they are *so* overlaid with gender. And yet the political and social contexts in which breast implants were developed have largely been overlooked.

Focusing on medical contexts alerts us to the ways in which our medical systems as a whole echo and reproduce theories of difference; the historical segregation of our minds and bodies into different disciplines influences our experiences in health and disease. Moreover, the objectifying gaze of scientific pathology has, in many cases, transmuted into psychological or social pathology. Thus having small breasts or being fat has become a psychiatric problem in the first instance and a social one in the second. With ever increasing specialities, moreover, it is inevitable that certain body parts are looked at through a specific and unwavering lens, as in the case of the scoliotic spine, which continues to be seen as a largely mechanical defect.

Much has changed since the time of Shakespeare, whose 'mortal coil' was borrowed for my title. We have moved away from a humoral age towards one of increasing specialization, and yet in some scientific discourses, like the rebranding of the gut as a site of intuition and feeling thanks to the emergence of the 'second brain,' we are arguably moving back to a material, if not a philosophical form of holism. Now, though, the body is being 'joined up' through gut flora, nerves, cellular memory, and hormones rather than through humours. Paradoxically, however, we are also more distancing of our bodies than ever before. The rise of cosmetic surgery—designer vaginas, pert breasts, tummy tucks, and lunchtime fillers—suggests we view our bodies as imperfect objects to be shaped at will, though there is often an unacknowledged psychological cost.[5] Women are also far more likely than men to sign up for cosmetic surgery, despite the fact that the specialism emerged as a response to war-time facial trauma.

One constant of the body in parts, then, is that those parts have always been regarded differently according to class, gender, and ethnicity. Presumptions made in Shakespeare's time about the sexual status of women, for instance, or about racial identity in the nineteenth century, drew on particular and historically sited forms of medico-scientific, social, and political knowledge. Another constant is the power of language to describe, produce, and naturalize difference. We see this in the way that the body is talked about, as well as the ways in which we organize our health-care systems around difference (between body parts and systems rather than between individuals). The body marks the limits of our selves; our skin is a symbolic and a physical boundary that separates us from others.[6] In this, the skin figures as a container, an envelope, a cover or a canvas. But that does not mean that we are separate from the world. We are social beings that communicate with others—not only through tongues as symbols and organs, but through social practices, behaviours and our very materiality. As the philosopher Maurice Merleau-Ponty puts it, our bodies are our 'way of being in the world'.[7] We are always aware of our bodies, though we might be lost in thought. Even language, which is normally opposed to the somatic, is simply confirmation of our physical, social existence.[8]

I have not talked much in this book about visual and material culture. There are important historiographical works that explore the representation of the body through drawings, paintings, sculptures, prints, advertisements and billboards.[9] I have been concerned principally by the ways language, and especially metaphor, shapes bodily experience. Metaphor is one of the main ways that we understand our bodies and our experiences in health and disease.[10] The metaphors that we use reflect the interests and parameters of the body politic (as a metaphor for the interests of the state) as well as the human body. In the above chapters we have seen the heart conceived as a pump and as an automaton, moving by clockwork and glowing with fire, obesity as waste, the spine as scaffolding as well as the breasts as pillows and the vagina as a deep pit or a toothed beast—a *vagina dentate*. The heart is a vessel that can be hot and full or cold and empty. The 'great chain of being' that stretched from the heavens to the rocks was itself a metaphor that maintained the social and political order as well as echoing somatic hierarchies; the monarch (the head) ruling over the lower social orders (from the groin to the feet).[11]

The heart has fluctuated in importance depending on the power source being reflected. For Galen, for whom it was the body's source of heat, the heart was a hearthstone. It fired up the body. It was also a 'root' through which the complex network of the body's veins and arteries spread and grew. William Harvey helped to rewrite the language of the heart when he popularized blood circulation. He dedicated his work to King Charles I just a few short years before civil war broke out, when the monarch had clear reasons for wanting to seem effortlessly and by nature in control of the kingdom. For Harvey the heart, like the king, was 'the foundation' of life, the 'sun of their microcosm, that upon which all growth depends, from which all power proceeds'.[12] In the writings of Descartes the heart was a pump, or better yet a combustion engine, the source of heat becoming a way to explain the rest of the automatic, machine-like body.

We continue to regard the heart as a pump, but we also use the language of the nerves and fibres of the eighteenth and nineteenth centuries, by which people can be 'highly strung' and 'nervous'. We view the emotional body as hydraulic; powerful feelings get pent up and have to be released slowly, like steam, to prevent us from 'blowing up' with rage.[13] Just as the pump metaphor was popularized at a time of mechanization, we have been talking about the brain as a computer since the 1970s when personal computers were first marketed; in the factory age it was, according to the English neurophysiologist Sir Charles Sherrington, an 'enchanted loom, where millions of flashing shuttles weave a dissolving pattern'.[14] Today we speak of recording experiences in our material brains, of filing them away in particular compartments in order to retrieve them at will. When we have traumatic or extreme mental experiences we have to 'process' them, rather as one would a series of computer files. Antonio Damasio has given this a physiological explanation, likening it to a 'movie-in-the-brain', part of the neural mechanisms by which we engage with and make sense of the world.[15] What this imagery fails to describe, as the sociologist Ian Burkitt notes, is how we creatively engage with those memory files; we interpret and change them and are not merely passive observers of our own visual movies.[16]

Metaphors are particularly prevalent around the female body and its menstrual cycles, which disrupt the stability of the body-as-factory; women's reproductive organs are supposed to be efficient baby-makers, giving rise to such medical terms as 'incompetent uterus' to describe a womb that does not operate as effectively as it ought. While menstruation is waste and 'failed' conception, menopause represents a broken down or outmoded machine.[17] As these politically laden examples suggest, metaphors are not simply a rhetorical flourish, or poetic turn of phrase. Our entire conceptual system—the way we think and act and feel—is metaphorical in nature.[18] Consider mind and body: they are both containers that we fill with things (ideas, food, images, emotions); it is virtually impossible to frame our experience or way of being in the world or to engage with others outside of these conceptual metaphors. Further examples are arguments, which have a competitive conceptual framework (we win, we lose, we concede a point or hit it home), or time, which is figured as a commodity that we save, lose or waste. Metaphors are more than mere descriptors: they inform and perpetuate the framework that they describe, and uphold ideas of difference.

In the future the brain may be redefined as an Internet search engine, a Google, taking over from the computer's archaic document retrieval system, just as the computer took over from the filing cabinet. There are a number of reasons why this might seem apposite. Firstly it reminds us of the importance of neuroplasticity; the widely held belief that entire brain structures and the brain itself can change and modify according to experience.[19] This kind of plasticity has even been identified in patterns of brain activity in users who are conducting Google searches.[20] More importantly, however, it is an important reminder of the social networks in which our brains develop. No less than the senses and the experience of touch, neural interconnectedness between self and the world is seen as necessary for mental and physical health. There are even established links between a lack of connectedness and neural degeneration, as found in studies on isolation and loneliness in dementia patients.[21]

We are social and somatic as well as psychological and individual beings, as shown by several chapters in this book. There are many more organs and body parts I might have discussed: the muscles, perhaps, or the hands or eyes. I chose the brain, the heart, and the guts because they are among the organs we associate most with feeling; the breasts and the female genitals because, as noted above, they are invested with particular discourses about ethnicity and gender. Tongues are not normally associated with sexual identity, though their functions are as invested with multiple meanings as any other part of our selves. I have concluded with the relationship between obesity and the gut because it brings together several of this book's main themes, including the relationship between mind and body, the stigmatisation of bodies that fall beyond the margins of the ideal and the ways knowledge might be possessed by the soma. Fatness is a visible reminder of how our boundaries move and develop with our physical selves. It has also become a signifier of the pathological—social as well as psychological and physiological—and attracts considerable social, medical and political debate.

How do we examine obesity without considering the physical, mental, spiritual and emotional aspects of a person? That is the question asked by integrated medicine, also known as integrative medicine in the United States, where it is better established. Integrated medicine allows for a more joined-up view of the self than allopathic or mainstream practice. The Royal College of Physicians of London has advocated its development in the UK because it combines complementary techniques with biomedicine, encouraging a focus on health and healing rather than on disease and treatment. It also puts patients' bodies back together with immaterial minds and even spirits. It seeks to overcome mind and body divisions and bring together patients and doctors in managing 'diet, exercise, quality of rest and sleep, and nature of relationships'.[22] If we added air and excretions to that list we would have the non-naturals of the humoral tradition.

Integrated medicine has an economic as well as a philosophical rationale. Modern biomedicine relies on technologies that are increasingly expensive. If some approved forms of complementary medicine successfully address chronic diseases then it takes the pressure off the NHS. It is also a trajectory that seems to acknowledge there are areas of somatic experience that do not have to be rationalized by mind; that habits and practices of the body, and forms of experience informed by the body, are particular *to* the body. The most obvious case is post-traumatic stress disorder (PTSD) when it is treated through body therapies that engage with the soma rather than the psyche.[23] An example of trialled NHS treatment is the Emotional Freedom Technique (EFT), also called 'tapping', which is believed to combat trauma.[24]

At a further remove, the possibility that emotional trauma is linked to 'cellular memory' (the belief that the body's cells can retain lived experiences, like the brain) is relatively new, and by no means mainstream. There are intriguing links, however, between the idea that feelings can be locked into the body's fibres—in the neurons of the heart and the gut, say—and the 'cross talk' between the brain and the rest of the body. Research into addiction, for instance, highlights cellular adaptation taking place as a response to drug taking. This phenomenon has been described as a form of 'cellular or molecular memory' that activates the same brain regions as other forms of embodied memory, from playing a musical instrument to eating a madeleine.[25] The notion of body memory is important: it suggests that somatic knowledge matters to our human identity and experiences as much as the mind.

The skin figures as a container, an envelope, a cover or a canvas. But that does not mean that we are separate from the world. We are social beings that communicate with others.

Integrated medicine brings its own languages to the understanding of bodies and health.[26] Conventional Western medicine is filled with fighting metaphors, with diseases as the invading enemy, with sporting metaphors about winning and losing as well as mechanistic metaphors that depict parts of the body as worn out and in need of repair.[27] One study suggests biomedical language stresses terms like 'credibility', 'legitimacy', 'scientifically proven', and 'efficiency' whereas integrative medicine is associated with terms like 'acceptance', 'empathy', 'healing', and 'nourishing' as well as 'cultural'.[28] There are as many ways of talking about medical practices, then, as there are of talking about the body itself.

Language is always changing, just like our interpretations of the body, but with the (re)birth of the gut-brain and the rise of cellular memory we are at a thought-provoking impasse. Might we see, in addition to new metaphors that describe a 'mind-in-body-in-society', probiotics used to treat bipolar disorder and 'gut feelings' being scrutinized by the kinds of measuring devices that dominated Victorian physiology? Epicurean philosophy, with its focus on the positivity of pleasure, may topple stoicism and its mind-over-matter severity.[29] Following your heart might be seen as an intelligent choice and an aid to reason.[30] Cosmetic surgery may become unnecessary in a world in which age is embraced, and wrinkles actively sought as hard-fought badges of honour. Perhaps this is a leap too far. But as embodied beings, we are surely more than the sum of our parts.

NOTES

Introduction. The Body in Parts

1 *Hamlet*, III. i. 58–71.
2 Hugh Grady, 'Introduction: Shakespeare and Modernity', in *Shakespeare and Modernity: Early Modern to Millennium*, ed. Hugh Grady (Hoboken, NJ: Taylor and Francis, 2013), 1–19, at 1; Hugh Grady, 'Renewing Modernity: Changing Contexts and Contents of a Nearly Invisible Concept', *Shakespeare Quarterly*, 50 (1999), 268–84, esp. 271.
3 Zhang Longxi, '"The Pale Cast of Thought": On the Dilemma of Thinking and Action', *New Literary History*, 45 (2014), 281–97; Brian Pearce, 'Hamlet, the Actor', *Shakespeare in Southern Africa*, 19 (2007), 63–9.
4 Gail Kern Paster, *Humoring the Body: Emotions and the Shakespearean Stage* (Chicago, IL: University of Chicago Press, 2004), 245.
5 *OED*, 'Mortal Coil'.
6 Margareta A. Sanner, 'Exchanging Spare Parts or Becoming a New Person? People's Attitudes toward Receiving and Donating Organs', *Social Science & Medicine*, 52 (2001), 1491–99.
7 Sarah Grogan, *Body Image: Understanding Body Dissatisfaction in Men, Women and Children* (London: Routledge, 2007).
8 Ryusuke Kagiki et al. 'Intracerebral Pain Processing in a Yoga Master who Claims not to Feel Pain during Meditation', *European Journal of Pain*, 9 (2005), 581–1.
9 David M. Friedman, *A Mind of its Own: A Cultural History of the Penis* (London: Simon and Schuster, 2008).
10 John Elsom, introduction in *Is Shakespeare Still Our Contemporary?*, ed. John Elsom (London: Routledge, 1989), 1.
11 Examples include Bryan S. Turner, *Routledge Handbook of Body Studies* (Abingdon-on-Thames: Routledge, 2012); Mariam Fraser and Monica Greco, *The Body: A Reader* (London: Routledge, 2004); Linda Kalof and William Bynum (eds), *A Cultural History of the Human Body*, 6 vols (Oxford: Berg, 2007); Jan Bremmer and Herman Roodenburg (eds), *A Cultural History of Gesture: From Antiquity to the Present Day* (Cambridge: Polity Press, 1991); Eugenia Paulicelli and Hazel Clark, *The Fabric of Cultures: Fashion, Identity and Globalization* (London: Routledge 2009).

12 Fay Bound Alberti, *Matters of the Heart: History, Medicine and Emotion* (Oxford: Oxford University Press, 2010); Joanna Bourke, *The Story of Pain: From Prayer to Painkillers* (Oxford: Oxford University Press, 2014); Javier Moscoso, *Pain: A Cultural History* (Basingstoke: Palgrave Macmillan, 2012); Andrew Shail and Gillian Howie, *Menstruation: A Cultural History* (Basingstoke: Palgrave Macmillan, 2005); David Howes, *A Cultural History of the Senses in the Modern Age, 1920–2000* (London: Bloomsbury Academic, 2014).

13 Donna Haraway, 'A Cyborg Manifesto: Science, Technology and Socialist-Feminism in the Late Twentieth Century', *Simians, Cyborgs and Women: The Reinvention of Nature* (London: Free Association, 1991), 149–82; Ana Delgado, 'Imagining High-Tech Bodies: Science Fiction and the Ethics of Enhancement', *Science Communication*, 34 (2012), 200–40.

14 Darin Weinberg, 'Social Constructionism and the Body', in Turner, *Routledge Handbook*, 144–56.

15 Classic texts include Michel Foucault, *Discipline and Punish: The Birth of the Prison* (London: Allen Lane, 1977) and Judith Butler, *Gender Trouble* (London: Routledge, 1990).

16 Caterina Albano, 'Visible Bodies: Cartography and Anatomy', in *Literature, Mapping, and the Politics of Space in Early Modern Britain*, ed. Andrew Gordon and Bernhard Klein (Cambridge: Cambridge University Press, 2001), 89–106. The exploration of a woman's body was frequently paralleled with that of new-found lands, as in the poems of John Donne: 'Your gown going off, such beauteous state reveals, | As when from flowery meads th'hill's shadow steals. | . . . | O my America, my new found land, | My kingdom, safeliest when with one man manned, | My Mine of precious stones, my empery, How blessed am I in this discovering thee.' 'Elegy 2: To his Mistress Going to Bed' in *John Donne: The Major Works*, edited with an introduction by John Carey (Oxford: Oxford University Press, 2008), 12, lines 13–30, with thanks to Miranda Bethell for this reference.

17 Antonio Damasio, *Descartes' Error: Emotion, Reason, and the Human Brain* (New York: Penguin Books, 2005), xii.

18 See Vanessa Heggie, 'Specialization without the Hospital: The Case of British Sports Medicine', *Medical History*, 54 (2010), 457–74.

19 Christopher Lawrence, *Medicine in the Making of Modern Britain, 1700–1920* (London and New York: Routledge, 1994), introduction.

20 Ahmad Hasheem et al., 'Medical Errors as a Result of Specialization', *Journal of Biomedical Informatics*, 36 (2003), 61–9.

21 Robert Peckham has made a similar association with crime. See 'Introduction: Pathologizing Crime, Criminalizing Disease', in *Disease and Crime: A History of Social Pathologies and the New Politics of Health*, ed. Robert Peckham (New York: Routledge, 2014), 1–20, esp. 13.

22 Charlotte Yeh, '"Nothing is Broken": For an Injured Doctor, Quality-Focused Care Misses the Mark', *Health Affairs*, 33 (2014), 1094–7; Maria Rosa Costanzo, 'The Physician Becomes the Patient: How My Breast Cancer Journey Taught Me to be a Better Doctor', *The Journal of Heart and Lung Transplantation*, 33 (2014), 1100–2; Sharon A. Hunt, 'But I'm a Doctor!' *The Journal of Heart and Lung Transplantation*, 33 (2014), 1103.

23 On the lack of communication training in orthopaedics in particular see Kristopher Lundine et al., 'Communication Skills Training in Orthopaedics', *The Journal of Bone & Joint Surgery*, 90 (2008), 1393–400.

24 Megan Tones et al., 'A Review of Quality of Life and Psychosocial issues in Scoliosis', *Spine*, 31 (2006), 3027–38; Chirag Patil et al., 'Inpatient Complications, Mortality and Discharge Disposition after Surgical Correction of Idiopathic Scoliosis: A National Perspective', *The Spine Journal*, 8 (2008), 904–10.

25 Kenneth J. Noonan et al., 'Long-Term Psychosocial Characteristics of Patients Treated for Idiopathic Scoliosis', *Journal of Pediatric Orthopaedics*, 17 (1997), 712–17.

26 Harold Ellis, *The Cambridge Illustrated History of Surgery* (Cambridge: Cambridge University Press, 2009), esp. ch. 10.

27 Stuart A. Green, 'Orthopaedic Surgeons: Inheritors of Tradition', *Clinical Orthopaedics and Related Research*, 363 (1999), 258–69; Roger Cooter, *Surgery and Society in Peace and War: Orthopaedics and the Organization of Modern Medicine, 1800–1948* (Basingstoke: Macmillan, 1993).

28 Russell C. Maulitz, *Morbid Appearances: The Anatomy of Pathology in the Early Nineteenth Century* (Cambridge: Cambridge University Press, 2002); Samuel J. M. M. Alberti, *Morbid Curiosities: Medical Museums in Nineteenth-Century Britain* (Oxford and New York: Oxford University Press, 2011).

29 Lynne Segal, *Straight Sex: The Politics of Pleasure* (London: Virago, 1994), 134.

30 Zbigniew Nawrat and Pawel Kostka, 'Polish Cardio-Robot "Robin Heart": System Description and Technical Evaluation', *The International Journal of Medical Robotics and Computer Assisted Surgery*, 2 (2006), 36–44.

31 Christiaan N. Barnard, 'Human Cardiac Transplantation: An Evaluation of the First Two Operations Performed at the Groote Schuur Hospital, Cape Town', *The American Journal of Cardiology*, 22 (1968), 584–96; Donald McRae et al., 'Every Second Counts: The Race to Transplant the First Human Heart', *Journal of Nuclear Medicine*, 48 (2007), 1571–2.

32 Galen, *On the Passions and Errors of the Soul*, trans. Paul W. Harkins (Columbus, OH: Ohio State University Press, 1963); Oswei Temkin, *Galenism: Rise and Decline of a Medical Philosophy* (Ithaca, NY: Cornell University Press, 1973), ch. 1.

33 Helkiah Crooke, *Mikrokosmographia: A Description of the Body of Man* (London, 1615); John Bernard Bamborough, *The Little World of Man* (London: Longmans, Green, 1952), ch. 1.

34 Helen King, 'Female Fluids in the Hippocratic Corpus: How Solid was the Humoral Body?', in *The Body in Balance: Humoral Medicines in Practice*, ed. Peregrine Horden and Elisabeth Hsu (New York: Berghahn Books, 2013), 25–52.
35 Temkin, *Galenism*, ch. 1.
36 Thomas Hobbes, *Leviathan*, ed. Richard Tuck (Düsseldorf and (Faks. d. Ausg.) London: Crooke, 1651; repr. Cambridge: Cambridge University Press, 1991), 53.
37 Thomas Wright, *Passions of the Minde in Generall* (London: VC for WB, 1601; facs. edn, Hildesheim, and New York: Olms, 1973), 109.
38 Thomas Wright, *Passions of the Minde*, 40.
39 Lelland Joseph Rather, 'Old and New Views of the Emotions and Bodily Changes: Wright and Harvey versus Descartes, James and Cannon', *Clio Medica*, 1 (1965), 1–25, at 4.
40 Bamborough, *The Little World of Man*, 64.
41 Thomas Wright, *Passions of the Minde*, 71–3.
42 Levinus Lemnius, *The Secret Miracles of Nature* (London: Streater, 1658), 274, cited in Bamborough, *Little World of Man*, 64.
43 *Hamlet*, IV. v. 76–7.
44 Carol Thomas Neely, *Distracted Subjects: Madness and Gender in Shakespeare and Early Modern Culture* (Ithaca, NY, and London: Cornell University Press, 2004).
45 Robert Burton, *The Anatomy Of Melancholy* (Oxford: John Lichfield and James Short for Henry Cripps, 1621) and see Ch. 4 of this book.
46 See Ch. 5 of this book.
47 Fred G. Barker, 'Phineas Among the Phrenologists: The American Crowbar Case and Nineteenth-Century Theories of Cerebral Localization, *Journal of Neurosurgery*, 82 (1995), 672–82.
48 Bound Alberti, *Matters of the Heart*, conclusion.
49 Philip C. Kolin, 'Blackness Made Visible: A Survey of *Othello* in Criticism, on Stage, and on Screen', in *Othello: New Critical Essays*, ed. Philip C. Kolin (Abingdon-on-Thames: Routledge, 2002), 1–88.
50 Roxann Wheeler, *The Complexion of Race: Categories of Difference in Eighteenth-Century British Culture* (Philadelphia, PA: University of Pennsylvania Press, 2000), 23.
51 George H. Bishop, 'The Skin as an Organ of Senses with Special Reference to the Itching Sensation', *Journal of Investigative Dermatology*, 11 (1948), 143–54.
52 Miriam Tilburg et al., 'Beyond "Pink It and Shrink It": Perceived Product Gender, Aesthetics, and Product Evaluation', *Psychology & Marketing* 32 (2015), 422–37, and Jackie Cook, 'Men's Magazines at the Millennium: New Spaces, New Selves', *Continuum: Journal of Media & Cultural Studies*, 14 (2000), 171–86.
53 Joni Hersch, 'Skin-Tone Effects among African Americans: Perceptions and Reality', *The American Economic Review*, 96 (2006), 251–5.
54 John J. Betancur (ed.), *Reinventing Race, Reinventing Racism* (Leiden and Boston, MA: Brill, 2013), 178.

55 Susan Gubar, *RaceChanges: White Skin, Black Face in American Culture* (New York and Oxford: Oxford University Press, 2000), 59.
56 Sadiah Qureshi, 'Displaying Sara Baartman, the "Hottentot Venus"', *History of Science*, 42 (2004), 233–57.
57 M. Nakayama, '[Histological Study on Aging Changes in the Human Tongue]', *Nihon Jibiinkoka Gakkai Kaiho*, 94 (1991), 541–55.
58 Harry T. Lawless and Hildegarde Heymann, *Sensory Evaluation of Food: Principles and Practices* (New York: Kluwer, 1999); Eric Schlosser, *Fast Food Nation: The Dark Side of the All-American Meal* (New York: Harper Perennial, 2005).
59 Francis Delpeuch, *Globesity: A Planet Out of Control?* (London: Routledge, 2013).
60 Gordon S. Bonham and Dwight B. Brock, 'The Relationship of Diabetes with Race, Sex, and Obesity', *The American Journal of Clinical Nutrition*, 41 (1985), 776–83; Barry M. Popkin, 'Does Global Obesity Represent a Global Public Health Challenge?' *The American Journal of Clinical Nutrition*, 93 (2011), 232–3.
61 Emily Q. Shults, 'Sharply Drawn Lines: An Examination of Title IX, Intersex, and Transgender', *Cardozo Journal of Law and Gender*, 12 (2005), 337.
62 Mark A. Hall and Carl E. Schneider, 'Patients as Consumers: Courts, Contracts, and the New Medical Marketplace', *Michigan Law Review*, 106 (2008), 643–89
63 Alice Kessler-Harris, 'The So-What Question', *Frontiers: A Journal of Women Studies*, 36 (2015), 12–16.
64 http://www.mentalhealth.org.uk/help-information/mental-health-statistics/ (accessed 1 September 2014).
65 Jonathan Shapiro, 'The NHS: The Story So Far (1948–2010)', *Clinical Medicine*, 10 (2010), 335–8.
66 Lifestyle diseases are responsible for perhaps two-thirds of the UK's NHS budget: 'The State of the Nation's Health', *The Times*, 25 May 2011, and WHO, Joint, and FAO Expert Consultation, 'Diet, Nutrition and the Prevention of Chronic Diseases', *WHO Technical Report Series*, 916 (2003), 1–60.
67 Bernd Moosmann and Christian Behl, 'Selenoprotein Synthesis and Side-Effects of Statins', *The Lancet*, 363 (2004), 892–4.
68 George L. Engel, 'The Clinical Application of the Biopsychosocial Model', *The American Journal of Psychiatry*, 137 (1980), 535–43; Niall McLaren, 'A Critical Review of the Biopsychosocial Model', *Australian and New Zealand Journal of Psychiatry*, 32 (1998), 86–92.
69 David Rakel (ed.), *Integrative Medicine* (Edinburgh: Elsevier Saunders, 2007).

Chapter 1. Getting it Straight: Spines, Scoliosis and the Hunchback King

1 *OED*, 'Spine, n.'. Definition of *spine* http://www.oxforddictionaries.com/definition/english/spine (accessed 26 November 2015).

2 *Richard III*, I. i. 14–23.

3 For a recent account see Alison Weir, *The Princes in the Tower* (London: Pimlico, 1997).

4 *On Ugliness*, ed. Umberto Eco, translated from the Italian by Alistair McEwen (London: MacLehose, 2011).

5 Anthony Synnott, 'Truth and Goodness, Mirrors and Masks, Part II: A Sociology of Beauty and the Face', *British Journal of Sociology*, 41 (1990), 55–76, at 55–6.

6 Victor Hugo, *The Hunchback of Notre-Dame*, abridged by Robin Waterfield (London: Puffin, 1996); *The Hunchback of Notre-Dame*, dir. Gary Trousdale and Kirk Wise, USA, Disney, 1996.

7 Nicolas Andry de Bois-Regard, *Orthopaedia, or, The Art of Preventing or Correcting Deformities in Children* (Birmingham, AL: The Classics of Medicine Library, 1980), introduction.

8 Sarah Knight and Mary Ann Lund, 'Richard Crookback', *Times Literary Supplement*, 6 February 2013, and see http://www.le.ac.uk/richardiii/ (accessed 1 March 2015).

9 Richard Buckley et al., '"The King in the Car Park": New Light on the Death and Burial of Richard III in the Grey Friars Church, Leicester, in 1485', *Antiquity*, 87 (2013), 519–38.

10 Isabel Tulloch, 'Richard III: A Study in Medical Misrepresentation', *Journal of the Royal Society of Medicine*, 102 (2009), 315–23.

11 Leviticus 21:16–23 (emphasis added).

12 Henri-Jacques Stiker, *A History of Disability*, trans. William Sayers (Ann Arbor, MI: University of Michigan Press, 1999), 24.

13 Krishna G. Seshadri, 'Hunches on Hunchbacks', *Indian Journal of Endocrinology and Metabolism*, 16 (2012), 292–4.

14 James Watkins, *Structure and Function of the Musculoskeletal System* (Leeds: Human Kinetics, 2010).

15 Andreas Vesalius, *De humani corporis fabrica* (Basle: Oporini, 1543).

16 Andrew Cunningham, *The Anatomical Renaissance: The Resurrection of the Anatomical Projects of the Ancients* (Aldershot: Ashgate, 1997).

17 http://vesalius.northwestern.edu/chapters/FA.1.16.html (accessed 1 March 2015).

18 Charles Singer, 'Galen's Elementary Discourse on Bones', *Proceedings of the Royal Society of Medicine*, 45 (1952), 767–76; Charles Mayo Goss and Elizabeth Goss Chodowski, 'On Bones For Beginners by Galen of Pergamon: A Translation with Commentary', *American Journal of Anatomy*, 169 (1984), 61–74.

19 Cited in Mayo Goss and Goss Chodowski, 'On Bones For Beginners', 61.

20 See introduction.

21 Koran cited in Lisa Sowle Cahill and Margaret A. Farley (eds), *Embodiment, Morality and Medicine* (Dordrecht, Boston, MA, and London: Kluwer Academic, 1995),

37; Dariusch Atighetchi, *Islamic Bioethics: Problems and Perspectives* (Dordrecht: Springer, 2006).

22 Genesis 2:21–2.

23 Scott A. Williams and Gabrielle A. Russo. 'Evolution of the Hominoid Vertebral Column: The Long and the Short of it', *Evolutionary Anthropology: Issues, News, and Reviews*, 24 (2015), 15–32.

24 Londa L. Schiebinger, 'Skeletons in the Closet: The First Illustrations of the Female Skeleton in Eighteenth-Century Anatomy', *Representations*, 14 (1986), 42–82; Londa L. Schiebinger, *The Mind Has No Sex: Women in the Origins of Modern Science* (Cambridge, MA: Harvard University Press, 1989).

25 Londa L. Schiebinger, 'The Philosopher's Beard: Women and Gender in Science', in *The Cambridge History of Science*, iv: *Eighteenth-Century Science*, ed. Roy Porter (Cambridge: Cambridge University Press, 2003), 184–210, esp. 199.

26 Catherine Gallagher and Thomas Walter Laqueur (eds), *The Making of the Modern Body: Sexuality and Society in the Nineteenth Century* (Berkeley, CA: University of California Press, 1986), xii.

27 Jaroslav Bruzek, 'A Method for Visual Determination of Sex, Using the Human Hip Bone', *American Journal of Physical Anthropology*, 117 (2002), 157–68.

28 Phillip L. Walker, 'Problems of Preservation and Sexism in Sexing: Some Lessons from Historical Collections for Palaeodemographers', *Age*, 18 (1995), 18–25, 20–1.

29 Walker, 'Problems of Preservation', 23.

30 Elias S. Vasiliadis et al., 'Historical Overview of Spinal Deformities in Ancient Greece', *Scoliosis*, 4 (2009), 4–6.

31 Vasiliadis et al., 'Historical Overview', 4.

32 Hidayet Sari et al., 'The Historical Development and Proof of Lumbar Traction Used in Physical Therapy', *Journal of Pharmacy and Pharmacology*, 2 (2014), 87–94.

33 Kathleen Y. Moen and Alf L. Nachemson, 'Treatment of Scoliosis: An Historical Perspective', *Spine*, 24 (1999), 2570–6.

34 Christa Lehnert-Schroth, *Three-Dimensional Treatment for Scoliosis: A Physiotherapeutic Method for Deformities of the Spine* (Palo Alto, CA: Martindale, 2000).

35 Brian V. Reamy and Joseph B. Slakey, 'Adolescent Idiopathic Scoliosis: Review and Current Concepts', *American Family Physician*, 64 (2001), 111–16.

36 James Tait Goodriche, 'History of Spine Surgery in the Ancient and Medieval Worlds', *Neurological Focus*, 16 (2004), 1–13.

37 Reginald S. Fayssouz, 'A History of Bracing for Idiopathic Scoliosis in North America', *Clinical Orthopaedics and Related Research*, 468 (2010), 654–64.

38 M-F. Weiner and J. R. Silver, 'Edward Harrison and the Treatment of Spinal Deformities in the Nineteenth Century', *The Journal of the Royal College of Physicians of Edinburgh*, 38 (2008), 265–71, at 266.

39 M-F. Weiner and J. R. Silver, 'Paralysis as a Result of Traction for the Treatment of Scoliosis: A Forgotten Lesson from History', *Spinal Cord*, 47 (2009), 429–34.
40 Weiner and Silver, 'Paralysis'.
41 Weiner and Silver, 'Edward Harrison'.
42 Edward Harrison, *Pathological and Practical Observations On Spinal Diseases: Illustrated with Cases and Engravings* (London: Underwood, 1827).
43 Harrison, *Pathological and Practical Observations*, 176.
44 Harrison, *Pathological and Practical Observations*, 177.
45 Harrison, *Pathological and Practical Observations*, 176.
46 A. N. Williams and J. Williams, '"Proper to the Duty of a Chirurgeon: Ambroise Paré and Sixteenth-Century Paediatric Surgery', *Journal of the Royal Society of Medicine*, 97 (2004), 446–9.
47 James W. Sayre, 'Lewis Albert Sayre', *Spine*, 20 (1995), 1091–6.
48 Andrew Chan et al., 'Review of Current Technologies and Methods Supplementing Brace Treatment in Adolescent Idiopathic Scoliosis', *Journal of Children's Orthopaedics*, 7 (2013), 309–16.
49 José M. Climent and José Sánchez, 'Impact of the Type of Brace on the Quality of Life of Adolescents with Spine Deformities', *Spine*, 24 (1999), 1903–8.
50 John E. Lonstein and Robert B. Winter, 'The Milwaukee Brace for the Treatment of Adolescent Idiopathic Scoliosis: A Review of One Thousand and Twenty Patients', *Journal of Bone & Joint Surgery*, 76 (1994), 1207–21.
51 Hans-Rudolf Weiss, 'Is there a Body of Evidence for the Treatment of Patients with Adolescent Idiopathic Scoliosis (AIS)?', *Scoliosis*, 2 (2007), 19–24.
52 Joshua D. Auerbach et al., 'Body Image in Patients with Adolescent Idiopathic Scoliosis', *Journal of Bone & Joint Surgery*, 96 (2014), e61.
53 Bareket W. Falk et al., 'Adolescent Idiopathic Scoliosis: The Possible Harm of Bracing and the Likely Benefit of Exercise', *The Spine Journal*, 15 (2015), 209–10.
54 K. Sansare et al., 'Short Communication: Early Victims of X-Rays: A Tribute and Current Perception', *Dentomaxillofacial Radiology*, 40 (2011), 123–5. On the transformative impact of X-rays along with atlases and films see Klaus Hentschel, *Visual Cultures in Science and Technology: A Comparative History* (New York: Oxford University Press, 2014), 290.
55 Otto Glasser, *Wilhelm Conrad Röntgen and the Early History of the Roentgen Rays* (San Francisco: Norman 1989).
56 Barron H. Lerner, 'The Perils of "X-Ray Vision": How Radiographic Images Have historically Influenced Perception', *Perspectives in Biology And Medicine*, 35 (1992), 382–97. Thanks to Sam Alberti for this reference.
57 Daniel A. Hoffman et al., 'Breast Cancer in Women with Scoliosis Exposed to Multiple Diagnostic X-Rays', *Journal of the National Cancer Institute*, 81 (1989), 1307–12.
58 Raymond R. Morrissy et al., 'Measurement of the Cobb Angle on Radiographs of Patients who have Scoliosis: Evaluation of Intrinsic Error', *Journal of Bone & Joint Surgery*, 72 (1990), 320–7.

59 J. E. Hans Pruijs et al., 'Variation in Cobb Angle Measurements in Scoliosis', *Skeletal Radiology*, 23 (1994), 517–20.
60 Peter V. Giannoudis et al., 'Bone Substitutes: An Update', *Injury*, 36 (2005), S20–S27.
61 Paul R. Harrington, 'Treatment of Scoliosis: Correction and Internal Fixation by Spine Instrumentation', *Journal of Bone and Joint Surgery*, 44 (1962), 591–610.
62 Alexander K. Meininger et al., 'Scapular Winging: An Update', *Journal of the American Academy of Orthopaedic Surgeons*, 19 (2011), 453–62.
63 Harshpal Singh and Allan D. Levi, 'Bone Graft and Bone Substitute Biology', in *Spine Surgery Basics*, ed. Vikas V. Patel et al. (Berlin: Springer, 2014), 147–52.
64 Robert Rawdon Wilson, 'Cyber(Body)parts: Prosthetic Consciousness', in *Cyberspace, Cyberbodies, Cyberpunk: Cultures of Technological Embodiment*, ed. Mike Featherstone and Roger Burrows (London: Sage, 1995), 239–60; Sean A. Hays (ed.), *Nanotechnology, the Brain and the Future* (Dordrecht: Springer, 2013).
65 Thomas Keller, 'Railway Spine Revisited: Traumatic Neurosis or Neurotrauma?', *Journal of the History of Medicine and Allied Sciences*, 50 (1995), 507–24, esp. 509, and the more nuanced approach by Rhodri Hayward, *The Transformation of the Psyche in British Primary Care, 1870–1970* (London: Bloomsbury, 2014), 13.
66 Robert French, *Robert Whytt, The Soul and Medicine* (London: The Wellcome Institute of Medical History, 1969), 90.
67 Hans Rudolf Weiss et al., 'Acupuncture in the Treatment of Scoliosis: A Single Blind Controlled Pilot Study', *Scoliosis* 3 (2008), 4.
68 Janet Richardson, 'What Patients Expect from Complementary Therapy: A Qualitative Study', *American Journal of Public Health*, 94 (2004), 1049–53.
69 Masafumi Machida et al., 'Melatonin: A Possible Role in Pathogenesis of Adolescent Idiopathic Scoliosis', *Spine*, 21 (1996), 1147–52; Harry K. W. Kim et al., 'Induction of SHP2 Deficiency in Chondrocytes Causes Severe Scoliosis and Kyphosis in Mice', *Spine*, 38 (2013), e1307–12; Tom P. C. Schlösser et al., 'How "Idiopathic" Is Adolescent Idiopathic Scoliosis? A Systematic Review on Associated Abnormalities', *PloS One*, 9 (2014), e97461.
70 http://www.thetimes.co.uk/tto/news/uk/royalfamily/article4393878.ece (accessed 28 March 2015).
71 http://www.telegraph.co.uk/history/9540207/Richard-III-skeleton-reveals-hunchback-king.html (accessed 1 December 2014).

Chapter 2. Beauty and the Breast: From Paraffin to PIP

1 Timmie Jean Lindsey, quoted in interview with the UK tabloid newspaper *The Daily Mail*: http://www.dailymail.co.uk/femail/article-484674/I-worlds-breast-job—endured-years-misery-says-Texan-great-grandmother.html (accessed 1 July 2014).
2 Marilyn Yalom, *A History of the Breast* (London: Pandora, 1998).
3 Iris Marion Young, 'Breasted Experience: The Look and the Feeling', *The Body in Medical Thought and Practice*, ed. Drew Leder (Dordrecht: Kluwer, 1992), 215–30.

4 Deborah Lupton, 'Femininity, Responsibility, and the Technological Imperative: Discourses on Breast Cancer in the Australian Press', *International Journal of Health Services*, 24 (1994), 73–90.

5 http://www.independent.co.uk/life-style/health-and-families/health-news/the-angelina-jolie-effect-her-mastectomy-revelation-doubled-nhs-breast-cancer-testing-referrals-9742074.html (accessed 1 December 2014).

6 D. Gareth R. Evans et al., 'The Angelina Jolie Effect: How High Celebrity Profile Can Have a Major Impact on Provision of Cancer-Related Services', *Breast Cancer Research*, 16 (2014), 442–7.

7 See also the award-winning poster for Breast Cancer Care: http://www.walesonline.co.uk/news/wales-news/cancer-survivor-jill-hindley-proudly-7987997 (accessed 12 December 2014).

8 Martha Grigg et al., 'Information for Women about the Safety of Silicone Breast Implants', *NIH Consensus Statement*, 15 (1997), 1–35.

9 Henri Wijsbek, 'The Pursuit of Beauty: The Enforcement of Aesthetics or a Freely Adopted Lifestyle?', *Journal of Medical Ethics*, 26 (2000), 454–8.

10 http://www.huffingtonpost.com/anthony-youn-md-facs/the-prespent-and-future-of_1_b_2864541.html (accessed 21 July 2014).

11 http://baaps.org.uk/about-us/press-releases/1833-britain-sucks (accessed 1 November 2013).

12 Robert A. Aronowitz, *Unnatural History: Breast Cancer and American Society* (Cambridge: Cambridge University Press, 2007); Alvin M. Cotlar et al., 'History of Surgery for Breast Cancer: Radical to the Sublime', *Journal of Current Surgery*, 60 (2003), 329–37; Sander L. Gilman, *Making the Body Beautiful: A Cultural History of Aesthetic Surgery* (Princeton, NJ: Princeton University Press, 1999); Elizabeth Haiken, *Venus Envy: A History of Cosmetic Surgery* (Baltimore, MD: Johns Hopkins University Press, 1997); Kathy Davis, *Reshaping the Female Body: The Dilemma of Cosmetic Surgery* (Abingdon-on-Thames: Routledge, 2013); Kathryn P. Morgan, 'Women and the Knife: Cosmetic Surgery and the Colonization of Women's Bodies', *Hypatia*, 6 (1991), 25–53.

13 Londa Shiebinger, 'Taxonomy for Human Beings', in *The Gendered Cyborg: A Reader*, ed. Gill Kirkup et al. (London and New York: Routledge, 2000), 11–37, 21; Willy Jansen and Catrien Notermans, 'Fluid Matters: Gendering Holy Hair and Holy Milk', in Dick Houtman and Birgit Meyer (eds), *Things: Religion and the Question of Materiality* (New York: Fordham University Press, 2012), 215–31, at 226.

14 Sir Astley Paston Cooper, *The Anatomy of the Breast* (London: Longman, Orme, Green, Browne and Longmans, 1840).

15 Cooper, *The Anatomy of the Breast*, 2.

16 Emily E. Stevens, 'A History of Infant Feeding', *The Journal of Perinatal Education*, 18 (2009), 32–9; Valerie Fildes, *Breasts, Bottles and Babies: A History of Infant Feeding* (Edinburgh: Edinburgh University Press, 1986).

17 Michele L. Crossley, 'Breastfeeding as a Moral Imperative: An Autoethnographic Study', *Feminism & Psychology*, 19 (2009), 71–87; Jane A. Scott and Tricia Mostyn, 'Women's Experiences of Breastfeeding in a Bottle-Feeding Culture', *Journal of Human Lactation*, 19 (2003), 270–7.

18 Cooper, *The Anatomy of the Breast*, 2.

19 Cooper, *The Anatomy of the Breast*, 18.

20 Sonnet 130, 3.

21 Adrian Desmond and James Moore, *Darwin's Sacred Cause: How a Hatred of Slavery Shaped Darwin's Views on Human Evolution* (Boston, MA: Houghton Mifflin Harcourt, 2009), 311.

22 e.g. Ann Olga Koloski-Ostrow and Claire L. Lyons, *Naked Truths: Women, Sex and Gender in Classical Art and Archaeology* (London and New York: Routledge, 2000), and Yalom, *History of the Breast*.

23 Marvin Rapaport, 'Silicone Injections Revisited', *Dermatologic Surgery*, 28 (2002), 594–5; Helen S. Edelman, 'Why is Dolly Crying? An Analysis of Silicone Breast Implants in America as an Example of Medicalization', *The Journal of Popular Culture*, 28 (1994), 19–32.

24 Shelly L. Grabe, et al., 'The Role of the Media in Body Image Concerns among Women: A Meta-Analysis of Experimental and Correlational Studies', *Psychological Bulletin*, 134 (2008), 460–76.

25 For an introduction to Susruta, see Ananda S. Chopra, 'Ayurveda', in *Medicine across Cultures: History and Practice of Medicine in Non-Western Cultures*, ed. Helaine Selin (Dordrecht and Boston, MA: Kluwer Academic Publishers, 2003), 75–84. See also Dominik Wujastyk, 'The Science of Medicine', in *The Blackwell Companion to Hinduism*, ed. Gavin Flood (Oxford: Blackwell, 2003), 393–409, and Rachel Berger, *Ayurveda Made Modern: Political Histories of Indigenous Medicine in North India, 1900–1955* (Basingstoke: Palgrave Macmillan, 2013).

26 John Lascaratos et al., 'Plastic Surgery of the Face in Byzantium in the Fourth Century', *Plastic and Reconstructive Surgery*, 102 (1998), 1274–80; Paolo Santoni-Rugiu and Philip J. Sykes, *A History of Plastic Surgery* (Berlin: Springer, 2007); D. J. Hauben, 'Sushruta Samhita (Sushruta's Collection) (800–600 BCE). Pioneers of Plastic Surgery', *Acta chirurgiae plasticae*, 26 (1983), 65–8.

27 Haiken, *Venus Envy*, 5.

28 Valentin Groebner, 'Losing Face, Saving Face: Noses and Honour in the Late Medieval Town', trans. Pamela Selwyn, *History Workshop Journal*, 40 (1995), 1–15. The mutilation of the nose has a long history as a form of punishment for adultery and fornication, amongst other crimes. In early modern culture it is also association with sexual infidelity and syphilis. See: Bruce T. Boehrer, 'Early Modern Syphilis', *Journal of the History of Sexuality*, 1 (1990), 197–214. The history of aesthetics around the nose is also the key theme for Sander L. Gilman in his work on cosmetic surgery and the 'body of the Jew'. See 'The Jewish Nose: Are Jews White? Or, The History of the Nose Job', in *The Other in Jewish Thought and History: Constructions of Jewish Culture and Identity*, ed.

Laurence J. Silberstein and Robert L. Cohn (New York: New York University Press, 1994), 364–401; Sander L. Gilman, *The Jew's Body* (New York: Routledge, 1991).

29 Daniel J. Hauben and Gijsbert J. Sonneveld, 'The Influence of War on the Development of Plastic Surgery', *Annals of Plastic Surgery*, 10 (1983), 65–9; James A. Chambers and Peter D. Ray, 'Achieving Growth and Excellence in Medicine: The Case History of Armed Conflict and Modern Reconstructive Surgery', *Annals of Plastic Surgery*, 63 (2009), 473–8; James A. Chambers et al., 'A Band of Surgeons, a Long Healing Line: Development of Craniofacial Surgery in Response to Armed Conflict', *Journal of Craniofacial Surgery*, 21 (2010), 991–7; Brian Morgan, 'War Surgery 1914 to 1918', *Journal of Plastic, Reconstructive & Aesthetic Surgery*, 66 (2013), 149.

30 Samuel J. M. M. Alberti (ed.), *War, Art and Surgery: The Work of Henry Tonks & Julia Midgley* (London: Royal College of Surgeons, 2014); Harold D. Gillies and D. Ralph Millard, *The Principles and Art of Plastic Surgery*, 2 vols (Boston, MA: Little, Brown, 1957).

31 Harold D. Gillies, *Plastic Surgery of the Face, Based on Selected Cases of War Injuries of the Face including Burns* (Oxford: Oxford University Press, 1920).

32 Murray C. Meikle, *Reconstructing Faces: The Art and Wartime Surgery of Gillies, Pickerill, McIndoe and Mowlem* (Dunedin: University of Otago Press, 2013).

33 M. G. Berry and D. M. Davies, 'Breast Augmentation, Part I: A Review of the Silicone Prosthesis', *Journal of Plastic, Reconstructive & Aesthetic Surgery*, 63 (2010), 1761–8, at 1761.

34 Robert M. Goldwyn, 'Vincenz Czerny and the Beginnings of Breast Reconstruction', *Plastic and Reconstructive Surgery*, 61 (1978), 673–81; Albert Losken and Maurice J. Jurkiewicz, 'History of Breast Reconstruction', *Breast Disease*, 15 (2002), 3–9.

35 David B. Sarwer et al., *Psychological Aspects of Reconstructive and Cosmetic Plastic Surgery: Clinical, Empirical, and Ethical Perspectives* (Philadelphia, PA: Lippincott Williams & Wilkins, 2005).

36 Søren Askegaard et al., 'The Body Consumed: Reflexivity and Cosmetic Surgery', *Psychology & Marketing*, 19 (2002), 793–812.

37 Sharon T. Phelan, 'Fads and Fashions: The Price Women Pay', *Primary Care Update for Ob/Gyns*, 9 (2002), 138–43.

38 Harold D. Gillies and Archibald H. McIndoe, 'The Technique of Mammoplasty in Conditions of Hypertrophy of the Breast', *Surgery, Gynecology and Obstetrics*, 68 (1939), 658–65; Nora Jacobson, 'The Socially Constructed Breast: Breast Implants and the Medical Construction of Need', *American Journal of Public Health*, 88 (1998), 1254–61; Antony F. Wallace, *The Progress of Plastic Surgery: An Introductory History* (Oxford: Meeuws, 1982); Ivo Pitanguy, 'Surgical Treatment of Breast Hypertrophy', *British Journal of Plastic Surgery*, 20 (1967), 78–85.

39 Elizabeth Haiken, 'Modern Miracles', in *Artificial Parts, Practical Lives: Modern Histories of Prosthetics*, ed. Katherine Ott et al. (New York: New York University Press, 2002), 171–98.

40 Virginia L. Blum, *Flesh Wounds: The Culture of Cosmetic Surgery* (Berkeley, CA: University of California Press, 2003), 14.
41 Cited in Haiken, 'Modern Miracles', 177.
42 Molly Haskell, *From Reverence to Rape: The Treatment of Women in the Movies* (Chicago, IL: University of Chicago Press, 1987), 235–52.
43 H. O. Bames, 'Correction of Abnormally Large or Small Breasts', *Southwestern Medicine*, January 1941, 11.
44 H. O. Bames, 'Plastic Reconstruction of the Anomalous Breast', *Revue de Chirurgie Structive*, July 1936, 294, and the discussion in Nora Jacobson, 'The Socially Constructed Breast', 1255.
45 Max Thorek, *Plastic Surgery of the Breast and Abdominal Wall* (Springfield, IL: Thomas, 1942).
46 H. O. Bames, 'Breast Malformations and a New Approach to the Problem of the Small Breast', *Plastic and Reconstructive Surgery*, 5 (1950), 499–506.
47 Bames, 'Breast Malformations', 499.
48 Gilman, *Making the Body Beautiful*, 242.
49 Courtney E. Martin, *Perfect Girls, Starving Daughters: The Frightening New Normalcy of Hating your Body* (New York: Simon and Schuster, 2007).
50 Alfred Adler, *The Education of Children*, trans. Eleanore Jensen and Friedrich Jensen (London: Allen and Unwin, 1930); Arthur C. Traub and J. Orbach, 'Psychophysical Studies of Body-Image, I: The Adjustable Body-Distorting Mirror', *Archives of General Psychiatry*, 11 (1964), 53–66; Anjan Chatterjee, 'Cosmetic Neurology and Cosmetic Surgery: Parallels, Predictions, and Challenges', *Cambridge Quarterly of Healthcare Ethics*, 16 (2007), 129–37; and Maxwell Maltz, *New Faces, New Futures: Rebuilding Character with Plastic Surgery* (New York: Smith, 1936).
51 Joy Leman, '"The Advice of a Real Friend": Codes of Intimacy and Oppression in Women's Magazines 1937–1955', *Women's Studies International Quarterly*, 3 (1980), 63–78; Naomi Wolf, *The Beauty Myth: How Images of Beauty are Used Against Women* (London: Random House, 1991); Helga Dittmar et al., 'Understanding the Impact of Thin Media Models on Women's Body-Focused Affect: The Roles of Thin-Ideal Internalization and Weight-Related Self-Discrepancy Activation in Experimental Exposure Effects', *Journal of Social and Clinical Psychology*, 28 (2009), 43–72, http://guilfordjournals.com/doi/abs/10.1521/jscp.2009.28.1.43 (accessed 28 November 2015).
52 Subsequently published as M. T. Edgerton and A. R. McClary, 'Augmentation Mammaplasty: Psychiatric Implications and Surgical Indications', *Plastic and Reconstructive Surgery*, 21 (1958), 279–305.
53 M. T. Edgerton et al., 'Augmentation Mammaplasty, II: Further Surgical and Psychiatric Evaluation', *Plastic and Reconstructive Surgery*, 27 (1961), 279–302.
54 James Barrett Brown et al., 'Polyvinyl and Solicone Compounds as Subcutaneous Prostheses: Laboratory and Clinical Investigation', *AMA Archives of Surgery*, 68

(1954), 744–51; William S. Kiskadden, 'Operations on Bosoms Dangerous', *Plastic and Reconstructive Surgery*, 15 (1955), 79–80.

55 Peter Conrad and Heather T. Jacobson, 'Enhancing Biology? Cosmetic Surgery and Breast Augmentation', in *Debating Biology: Sociological Reflections on Health, Medicine and Society*, ed. Gillian Bendelow et al. (London: Routledge, 2003), 223–35.

56 Berry and Davies, 'Breast Augmentation'; Marlene Johnson, 'Breast Implants: History, Safety, and Imaging', *Radiologic Technology*, 84 (2013), 439M–520M; Laura Miller, 'Mammary Mania in Japan', *Positions: East Asia Cultures Critique*, 11 (2003), 271–300.

57 http://www.theguardian.com/lifeandstyle/2008/may/03/healthandwellbeing.health (accessed 27 November 2015).

58 Thomas D. Cronin and Frank J. Gerow, 'Augmentation Mammoplasty: A New "Natural Feel" Prosthesis', *Transactions of the Third International Congress of Plastic Surgery*, 66 (1964), 41–9; Nora Jacobson, 'The Socially Constructed Breast', 1256; Susan M. Zimmerman, *Silicone Survivors: Women's Experiences with Breast Implants* (Temple University Press: Philadelphia, PA, 1998), ch. 2; Marsha L. Vanderford and David H. Smith, *The Silicone Breast Implant Story: Communication and Uncertainty* (Abingdon-on-Thames: Routledge, 2013), 11.

59 Mary Jacobus et al. (eds), *Body/Politics: Women and the Discourses of Science* (New York and London: Routledge, 1989); Londa Schiebinger, *Nature's Body: Gender in the Making of Modern Science* (New Brunswick, NJ: Rutgers University Press, 1993).

60 Thomas Cronin and Roger Greenberg, 'Our Experiences with the Silastic Gel Breast Prosthesis', *Plastic and Reconstructive Surgery*, 46 (1970), 1–7, at 1.

61 Interview with Thomas Biggs in http://cosmeticsurgerytimes.modernmedicine.com/cosmetic-surgery-times/news/modernmedicine/modern-medicine-feature-articles/modern-breast-implants-e (accessed 1 December 2014).

62 Grigg et al., 'Information for Women about the Safety of Silicone Breast Implants'.

63 Ralph Blocksma and Silas Braley, 'The Silicones in Plastic Surgery', *Plastic and Reconstructive Surgery*, 35 (1965), 366–70.

64 Interview with Thomas Biggs in http://cosmeticsurgerytimes.modernmedicine.com/cosmetic-surgery-times/news/modernmedicine/modern-medicine-feature-articles/modern-breast-implants-e (accessed 1 December 2014).

65 http://www.ocregister.com/articles/implants-354332-breast-silicone.html (accessed 1 November 2014) and http://www.dailymail.co.uk/femail/article-484674/I-worlds-breast-job—endured-years-misery-says-Texan-great-grandmother.html (accessed 1 July 2014).

66 http://www.dailymail.co.uk/femail/article-484674/I-worlds-breast-job—endured-years-misery-says-Texan-great-grandmother.html (accessed 1 July 2014).

67 Nora Jacobson, 'The Socially Constructed Breast', 1256.

68 Marcia Angell, 'Evaluating the Health Risks of Breast Implants: The Interplay of Medical Science, the Law, and Public Opinion', *New England Journal of Medicine*, 334 (1996), 1513–18, on 1513–14.
69 Vanderford and Smith, *The Silicone Breast Implant Story*, 11.
70 Arshad R. Muzaffar and Rod J. Rohrich, 'The Silicone Gel-Filled Breast Implant Controversy: An Update', *Plastic and Reconstructive Surgery*, 109 (2002), 742–8, at 743; Nanette D. DeBruhl, et al., 'Silicone Breast Implants: US Evaluation', *Radiology*, 189 (1993), 95–8; Julie M. Spanbauer, 'Breast Implants as Beauty Ritual: Woman's Sceptre and Prison', *Yale Journal of Law & Feminism*, 9 (1997), 157–206.
71 Thomas M. Sinclair et al., 'Biodegradation of the Polyurethane Foam Covering of Breast Implants', *Plastic and Reconstructive Surgery*, 2 (1993), 1003–13.
72 Susan Bartlett Foote, 'Loop and Loopholes: Hazardous Device Regulation under the 1976 Medical Device Amendments to the Food, Drug and Cosmetic Act', *Ecology Law Quarterly*, 7 (1978), 101–35.
73 Medical Device Amendments of 1976, P. L. 94-295 coded at 21 U.S.C. Section 360.
74 Foote, 'Loop and Loopholes'; Susan Bartlett Foote et al., 'The Impact of Public Policy on Medical Device Innovation: A Case of Polyintervention', in *The Changing Economics of Medical Innovation in Medicine*, Annetine C. Gelijns and Ethan A. Halm (Washington DC: National Academy Press, 1991), 69–88.
75 Robin E. Stombler, 'Breast Implants and the FDA: Past, Present, and Future', *Plastic Surgical Nursing*, 13 (1993), 185–7.
76 Yasuo Kumagai et al., 'Scleroderma after Cosmetic Surgery: Four Cases of Human Adjuvant Disease', *Arthritis & Rheumatology*, 22 (1979), 532–7; Yasuo Kumagai et al., 'Clinical Spectrum of Connective Tissue Disease after Cosmetic Surgery: Observation on Eighteen Patients and a Review of the Japanese Literature', *Arthritis & Rheumatology*, 27 (1984), 1–12.
77 Angell, 'Evaluating the Health Risks', 1513–14; Sheila Jasanoff, 'Science and the Statistical Victim Modernizing Knowledge in Breast Implant Litigation', *Social Studies of Science*, 32 (2002), 37–69.
78 Jasanoff, 'Science and the Statistical Victim'.
79 Angela Powers and Julie L. Andsager, 'How Newspapers Framed Breast Implants in the 1990s', *Journalism & Mass Communication Quarterly*, 76 (1999), 551–64.
80 *Face to Face with Connie Chung*, CBS Broadcasting, 10 December 1990.
81 Angell, 'Evaluating the Health Risks', 1513–14; Charlotte Allen, 'Jurisprudence of Breasts', *Stanford Law and Policy Review*, 5 (1993), 83–90; Zoe Panarites, 'Breast Implants: Choices Women Thought They Made', *New York School Journal of Human Rights*, 11 (1993), 163–204.
82 Angell, 'Evaluating the Health Risks', 1514.
83 Vanderford and Smith, *The Silicone Breast Implant Story*, 12–13.
84 Britta Ostermeyer Shoaib and Bernard M. Patten, 'A Motor Neuron Disease Syndrome in Silicone Breast Implant Recipients', *Journal of Occupational Medicine*

and Toxicology, 4 (1995), 155–63; Britta Ostermeyer Shoaib and Bernard M. Patten, 'Human Adjuvant Disease: Presentation as a Multiple Sclerosis-Like Syndrome', *Southern Medical Journal*, 89 (1996), 179–88.

85 Tonja E. Olive, 'Vilification Stories: The Fall of Dow Corning', in Vanderford and Smith, *The Silicone Breast Implant Story*, 148–76, 150.

86 Rebecca Weisman, 'Reforms in Medical Device Regulation: An Examination of the Silicone Gel Breast Implant Debacle', *Golden Gate University Law Review*, 23 (1993), 973–1000.

87 Ralph R. Cook et al., 'The Breast Implant Controversy', *Arthritis & Rheumatism*, 37 (1994), 153–7; Paul A. Argenti, 'How Should Reputations be Managed in Good Times and Bad Times? Dow Corning's Breast Implant Controversy: Managing Reputation in the Face of "Junk Science" ', *Corporate Reputation Review*, 1 (1997), 126–31.

88 David A. Kessler, 'The Basis of the FDA's Decision on Breast Implants', *New England Journal of Medicine*, 326 (1992), 1713–15.

89 Kessler, 'The Basis of the FDA's Decision', 1713–14.

90 Kessler, 'The Basis of the FDA's Decision', 1714.

91 Angell, *Science on Trial: The Clash of Medical Evidence and the Law in the Breast Implant Case* (London: Norton, 1997).

92 Joseph K. McLaughlin et al., 'The Safety of Silicone Gel-Filled Breast Implants: A Review of the Epidemiologic Evidence', *Annals of Plastic Surgery*, 59 (2007), 569–80, abstract; Peter Tugwell et al., 'Do Silicone Breast Implants Cause Rheumatologic Disorders? A Systematic Review for a Court-Appointed National Science Panel', *Arthritis & Rheumatism*, 44 (2001), 2477–84.

93 Stuart Bondurant, et al., *Safety of Silicone Breast Implants: Report of the Committee on the Safety of Silicone Breast Implants* (Washington DC: Institute of Medicine, National Academy Press, 1999).

94 Bondurant, et al., *Safety of Silicone Breast Implants*, 7.

95 Bondurant, et al., *Safety of Silicone Breast Implants*, 7.

96 Bondurant, et al., *Safety of Silicone Breast Implants*, 229.

97 David B. Sarwer et al., 'Cosmetic Breast Augmentation Surgery: A Critical Overview', *Journal of Women's Health & Gender-Based Medicine*, 9 (2000), 843–56.

98 http://www.fda.gov/ForConsumers/ConsumerUpdates/ucm259825.htm (accessed 17 November 2014).

99 Adrian O'Dowd, 'Around 1000 Women with Private Sector PIP Implants Seek NHS Help', *BMJ*, 344 (2012), e972; Adrian O'Dowd, 'Government Puts Pressure on Private Sector to Pay for Removal of PIP Breast Implants', *BMJ*, 344 (2012), e249; Ingrid Torjesen, 'Hundreds of Thousands of Pounds of NHS Funds Have Been Spent on Care of Private Patients with PIP Implants', *BMJ*, 344 (2012), e1259.

100 http://www.bbc.co.uk/news/health-16749773 (accessed 7 September 2014).

101 M. G. Berry and Jan J. Stanek, 'The PIP Mammary Prosthesis: A Product Recall Study', *Journal of Plastic, Reconstructive & Aesthetic Surgery*, 65 (2012), 697–704.
102 Paul Benkimoun, 'Founder of PIP Breast Implant Company Gets Four-Year Prison Sentence', *BMJ*, 347 (2013), f7528.
103 http://www.liverpoolecho.co.uk/news/liverpool-news/pip-breast-implants-killing-me-3354114 (accessed 21 July 2014).
104 Deborah Cohen and Matthew Billingsley, 'Europeans are Left to their own Devices', *BMJ*, 342 (2011), d2748.
105 Review of the Regulation of Cosmetic Interventions, 2013: https://www.gov.uk/government/uploads/system/uploads/attachment_data/file/192028/Review_of_the_Regulation_of_Cosmetic_Interventions.pdf (accessed 1 October 2014).
106 http://www.theguardian.com/world/2013/oct/02/french-breast-implant-scandal-france (accessed 27 November 2015).
107 Richard Horton, 'Offline: A Serious Regulatory Failure, With Urgent Implications', *The Lancet*, 379 (2012), 106.
108 Shaheel Chummun and Neil R. McLean, 'Poly Implant Prothèse (PIP) Breast Implants: Our Experience', *The Surgeon*, 11 (2013), 241–5.
109 Chummun and McLean, 'Poly Implant Prothèse', Introduction.
110 Poly Implant Prothese (PIP) Breast Implants: Final Report of the Working Group, 18 June 2012, Department of Health, NHS Medical Directorate.
111 Breast Implants: Final Report of the Working Group, recommendation 31.
112 http://www.harleymedical.co.uk (accessed 2 October 2013).
113 http://www.thesun.co.uk/sol/homepage/news/3875577/Test-has-assessed-the-best-breasts.html (accessed 1 July 2014).
114 Barnaby J. Dixson et al., 'Men's Preferences for Women's Breast Morphology in New Zealand, Samoa, and Papua New Guinea', *Archives of Sexual Behavior*, 40 (2011), 1271–9.
115 Behnaz Schofield, 'The Role of Consent and Individual Autonomy in the PIP Breast Implant Scandal', *Public Health Ethics*, 6 (2013), 220–3.
116 Bruce Keogh, *Review of the Regulation of Cosmetic Interventions* (London: Department of Health, 2013).
117 Keogh, *Review*, foreword.
118 http://www.cc4plasticsurgery.com/cosmetic/breast/breast-reduction/ (accessed 14 July 2014) and http://www.harleymedical.co.uk/cosmetic-surgery-for-women/breast-surgery/breast-surgery-overview (accessed 14 July 2014).
119 Marsha L. Richins, 'Social Comparison and the Idealized Images of Advertising', *Journal of Consumer Research*, 18 (1991), 71–83.
120 http://www.harleymedical.co.uk/cosmetic-surgery-for-women/breast-surgery/breast-surgery-overview (accessed 14 July 2014).
121 Franklin G. Miller et al., 'Cosmetic Surgery and the Internal Morality of Medicine', *Cambridge Quarterly of Healthcare Ethics*, 9 (2000), 353–64.

122 https://www.gov.uk/government/publications/review-of-the-regulation-of-cosmetic-interventions (accessed 14 July 2014).
123 http://baaps.org.uk/about-us/press-releases/1624-one-in-five-unsuitable-for-cosmetic-surgery-patients-dangerously-misinformed-by-salespeople (accessed 14 July 2014).
124 Keogh, *Review*, 3.3.
125 Melanie Latham, '"If it Ain't Broke, Don't Fix it?" Scandals, "Risk", and Cosmetic Surgery Regulation in the UK and France', *Medical Law Review*, 22 (2014), 384–408.
126 National Care Standards Commission, Report to the Chief Medical Officer for England on the Findings of Inspectors of Private and Cosmetic Surgery Establishments in Central London during March/April 2003, June 2003, discussed by Latham, '"If it Ain't Broke, Don't Fix it?"'.
127 J. J. B. Tinkler et al., *Evidence for an Association Between the Implantation of Silicones and Connective Tissue Disease* (UK Department of Health, Medical Services Directorate, 1993).
128 Latham, '"If it Ain't Broke, Don't Fix it?"'; correspondence between Mr Brian Morgan and the Senior Medical Officer at the Medical Services Directorate, Department of Health, dated 24 November 1992: BAPRAS archive at the Royal College of Surgeons of England.
129 HC Deb, 17 July 2012, c856.
130 http://www.express.co.uk/finance/city/412963/Silicon-scare-helps-plump-up-sales-at-UK-s-Nagor (accessed 2 October 2013).
131 https://www.rcseng.ac.uk/news/surgeons-publish-landmark-standards-for-cosmetic-practice#.VL5DTVrz1Ec (accessed 20 January 2013).
132 https://www.rcseng.ac.uk/news/surgeons-publish-landmark-standards-for-cosmetic-practice#.VL5DTVrz1Ec (accessed 20 January 2013).
133 http://www.theguardian.com/lifeandstyle/2008/may/03/healthandwellbeing.health (accessed 27 November 2015).
134 http://www.theguardian.com/lifeandstyle/2008/may/03/healthandwellbeing.health (accessed 27 November 2015).
135 Keogh, *Review*, 3.3.
136 David B. Sarwer et al., 'An Investigation of Changes in Body Image Following Cosmetic Surgery', *Plastic and Reconstructive Surgery*, 109 (2002), 363–9.
137 Melanie Latham, 'The Shape of Things to Come', *Medical Law Review*, 16 (2008), 437–57.
138 Vanderford and Smith, *The Silicone Breast Implant Story*, introduction.
139 Morgan, 'Women and the Knife'.
140 Cressida Hayes and Meredith Jones (eds), *Cosmetic Surgery: A Feminist Primer* (Farnham: Ashgate, 2009).
141 Davis, *Reshaping the Female Body*.

142 Anne Marie Balsamo, *Technologies of the Gendered Body: Reading Cyborg Women* (Durham, NC: Duke University Press, 1996), 78.
143 Annie Potts, 'The Mark of the Beast: Inscribing 'Animality' through Extreme Body Modification', in *Knowing Animals*, ed. Laurence Simmons and Philip Armstrong (Leiden and Boston, MA: Brill, 2007), 131–54.
144 Clare Chambers, 'Are Breast Implants Better than Female Genital Mutilation? Autonomy, Gender Equality and Nussbaum's Political Liberalism', *Critical Review of International Social and Political Philosophy*, 7 (2004), 1–33.
145 David B. Sarwer et al., 'Body Image Concerns of Breast Augmentation Patients', *Plastic and Reconstructive Surgery*, 112 (2003), 83–90.
146 Loren Lipworth et al., 'Excess Mortality from Suicide and Other External Causes of Death among Women with Cosmetic Breast Implants', *Annals of Plastic Surgery*, 59 (2007), 119–23.
147 Rachel Nowak, 'When Looks Can Kill', 2574, *New Scientist* (2006), 18; V. C. M. Koot et al., 'Total and Cause-Specific Mortality among Swedish Women with Cosmetic Breast Implants', *BMJ*, 326 (2003), 527.
148 Susie Orbach, *Bodies* (London: Profile Books, 2009), 22–3.
149 Vanderford and Smith, *The Silicone Breast Implant Story*, 8–9; Elaine Showalter, *The Female Malady: Women, Madness and Culture in England, 1830–1980* (New York: Pantheon, 1986).
150 Vanderford and Smith, *The Silicone Breast Implant Story*, 9; Nance Cunningham, 'Power, the Meaning of Subjectivity and Public Education about Fibromyalgia', *Journal of Philosophy and History of Education*, 59 (2009), 186–90; Kathleen Kendall, 'Masking Violence Against Women: The Case of PMS', *Canadian Woman Studies*, 12 (1991), 17–20.
151 http://surgicalcareers.rcseng.ac.uk/wins/statistics (accessed 22 January 2015).
152 http://www.barereality.net/about (accessed 9 September 2014).
153 Susan Bordo, *Unbearable Weight: Feminism, Western Culture, and the Body*, Tenth Anniversary Edition (Berkeley, CA: University of California Press, 2003), 139.

Chapter 3. 'Country Matters': The Language and Politics of Female Genitalia

1 Gabriel Egan, *Shakespeare* (Edinburgh: Edinburgh University Press, 2007), 37.
2 Ali Reza Ghanooni, 'Sexual Pun: A Case Study of Shakespeare's *Romeo and Juliet*', *Cross-Cultural Communication*, 8 (2012), 91–100. See also Pauline Kiernan, *Filthy Shakespeare: Shakespeare's Most Outrageous Sexual Puns* (London: Quercus, 2006); Philip Armstrong, 'Hamlet and Ophelia: Watching Hamlet Watching', in *Alternative Shakespeares*, 2 vols (London: Routledge, 2003), Vol. 2, 216–34; Gordon Williams, *Shakespeare, Sex and the Print Revolution* (London: Athlone, 1996); Stanley Wells, *Shakespeare, Sex and Love* (Oxford: Oxford University

Press, 2010); Stanley Wells, *Looking for Sex in Shakespeare* (Cambridge: Cambridge University Press, 2004).

3 Theodora A. Jankowski, *Women in Power in the Early Modern Drama* (Urbana, IL: University of Illinois Press, 1992), 5.

4 Thomas Laqueur, *Making Sex: Body and Gender from the Greeks to Freud* (Cambridge, MA: Harvard University Press, 1990), 25; Stephen Greenblatt, *Shakespearean Negotiations: The Circulation of Social Energy in Renaissance England* (Oxford: Clarendon Press, 1990), ch. 3.

5 Winifred Schleiner, 'Early Modern Controversies about the One-Sex Model', *Renaissance Quarterly*, 53 (2000), 180–91; Janet Adelman, 'Making Defect Perfection: Shakespeare and the One-Sex Model', in *Enacting Gender on the English Renaissance Stage*, ed. Viviana Comensoli and Anne Russell (Urbana, IL: University of Illinois Press, 1999), 23–52.

6 Carol Chillington Rutter, *Enter the Body: Women and Representation on Shakespeare's Stage* (Hoboken, NJ: Taylor and Francis, 2002); Marguerite A. Tassi, *Women and Revenge in Shakespeare: Gender, Genre and Ethics* (Selinsgrove, PA: Susquehanna University Press, 2011).

7 See Richard A. Levin, 'Shakespeare's Sonnets 153 and 154', *The Explicator*, 53 (1994), 11–14; Peter L. Rudnytsky, 'The "Darke and Vicious Place": The Dread of the Vagina in King Lear', *Modern Philology*, 96 (1999), 291–311; David Carnegie, 'Early Modern Plays and Performance', *Huntington Library Quarterly*, 67 (2004), 437–56; Theodora A. Janowski, 'Hymeneal Blood, Interchangeable Women and the Early Modern Marriage Economy in *Measure for Measure* and *All's Well that Ends Well*', in *A Companion to Shakespeare's Works*, ed. Richard Dutton and Jean E. Howard, 4 vols (Malden, MA, and Oxford: Blackwell, 2003), Vol. 4, 89–105.

8 Eve Ensler, *The Vagina Monologues* (New York: Villard, 2001), preface.

9 Virginia Braun and Celia Kitzinger, 'Telling it Straight? Dictionary Definitions of Women's Genitals', *Journal of Sociolinguistics*, 5 (2001), 214–32, at 215.

10 Lesbian, Gay, Bisexual, Transgender and others.

11 Christine M. Cooper, 'Worrying About Vaginas: Feminism and Eve Ensler's *The Vagina Monologues*', *Signs: Journal of Women in Culture and Society*, 32 (2007), 727–58; Virginia Braun and Sue Wilkinson, 'Vagina Equals Woman? On Genitals and Gendered Identity', *Women's Studies International Forum*, 28 (2005), 509–22.

12 Sherry Ortner, 'Is Female to Male as Nature is to Culture?' *Feminist Studies*, 1 (1972), 5–31; R. A. Sydie, *Natural Women, Cultured Men: A Feminist Perspective on Sociological Theory* (Vancouver: UBC Press, 1994); Karen Harvey, *Reading Sex in the Eighteenth Century: Bodies and Gender in English Erotic Culture* (Cambridge: Cambridge University Press, 2008), 106.

13 Georgianna Ziegler, 'My Lady's Chamber: Female Space, Female Chastity in Shakespeare', *Textual Practice*, 4 (1990), 73–90; Joanna Groot and Sue Morgan, 'Beyond the "Religious Turn"? Past, Present and Future Perspectives in Gender History', *Gender & History*, 25 (2013), 395–422; Jeffrey Weeks, *Sex, Politics and*

Society: The Regulation of Sexuality since 1800 (Harlow: Pearson Education, 2012), 102–3; Peter C. Engelman, *A History of the Birth Control Movement in America* (Santa Barbara, CA: Praeger, 2011).

14 Roy F. Baumeister and Jean M. Twenge. 'Cultural Suppression of Female Sexuality', *Review of General Psychology*, 6 (2002), 166–203; Daniel Bell, 'The East Asian Challenge to Human Rights: Reflections on an East–West Dialogue', *Human Rights Quarterly*, 18 (1996), 641–67.

15 Michelle L. Hammers, 'Talking About "Down There": The Politics of Publicizing the Female Body through The Vagina Monologues', *Women's Studies in Communication*, 29 (2006), 220–43; Emma L. E. Rees, *The Vagina: A Literary and Cultural History* (London: Bloomsbury, 2013), 33.

16 https://www.gov.uk/national-curriculum/other-compulsory-subjects (accessed 26 February 2015).

17 Germaine Greer, *The Whole Woman* (New York: Knopf, 1999), 39.

18 Roy J. Levin, 'Can the Controversy about the Putative Role of the Human Female Orgasm in Sperm Transport be Settled with our Current Physiological Knowledge of Coitus?' *The Journal of Sexual Medicine*, 8 (2011), 1566–78.

19 Elisabeth A. Lloyd, *The Case of the Female Orgasm: Bias in the Science of Evolution* (Cambridge, MA: Harvard University Press, 2005).

20 Liza Picard, *Restoration London: Everyday Life in the 1660s* (London: Phoenix, 2003).

21 Jane Sharp, *The Midwives Book, or, The Whole Art of Midwifry Discovered*, ed. Elaine Hobby (New York: Oxford University Press, 1999), 80; Michele Goodwin, 'Law's Limits: Regulating Statutory Rape Law', *Wisconsin Law Review* (2013), 13–23, at 14.

22 Sharp, *The Midwives Book*, xxxi.

23 Gérard. Zwang, 'Vulvar Reconstruction. The Exploitation of an Ignorance', *Sexologies*, 20 (2011), 81–7.

24 Vanessa R. Schick et al., 'E vulva lution: The Portrayal of Women's External Genitalia and Physique Across Time and the Current *Barbie* Doll Ideals', *Journal of Sex Research*, 48 (2011), 74–81.

25 See for instance Ada Borkenhagen and Heribert Kentenich, 'Labia Reduction: The Newest Trend in Cosmetic Genitoplasty, An Overview', *Geburtshilfe und Frauenheilkunde*, 69 (2009), 19–23, and David Veale et al., 'A Comparison of Risk Factors for Women Seeking Labiaplasty Compared to Those Not Seeking Labiaplasty', *Body Image*, 11 (2014), 57–62.

26 http://www.greatwallofvagina.co.uk/home (accessed 12 September 2014).

27 Hugh Aldersey-Williams, *Anatomies: The Human Body, its Parts and the Stories they Tell* (London: Penguin, 2012), xxii.

28 http://www.who.int/mediacentre/factsheets/fs241/en/index.html (accessed 22 March 2015); Sara Rodrigues, 'From Vaginal Exception to Exceptional Vagina: The Biopolitics of Female Genital Cosmetic Surgery', *Sexualities*, 15 (2012), 778–95.

29 Marika Tiggemann and Suzanna Hodgson, 'The Hairlessness Norm Extended: Reasons for and Predictors of Women's Hair Removal at Different Body Sites', *Sex Roles*, 59 (2008), 889–97.

30 Susan A. Basow, 'The Hairless Ideal: Women and their Body Hair', *Psychology of Women Quarterly*, 15 (1991), 83–96.

31 Magdala Peixoto Labre, 'The Brazilian Wax: New Hairlessness Norm for Women?' *Journal of Communication Inquiry*, 26 (2002), 113–32.

32 Merran Toerien and Sue Wilkinson, 'Exploring the Depilation Norm: A Qualitative Questionnaire Study of Women's Body Hair Removal', *Qualitative Research in Psychology*, 1 (2004), 69–92.

33 Christopher Hals Gylseth and Lars O. Toverud, *Julia Pastrana: The Tragic Story Of The Victorian Ape Woman*, trans. Donald Tumasonis (Stroud: Sutton Publishing, 2004).

34 Janet Browne and Sharon Messenger, 'Victorian Spectacle: Julia Pastrana, the Bearded and Hairy Female', *Endeavour*, 27 (2003), 155–59, 155.

35 Mark A. Katrizky, 'Literary Anthropologies and Pedro González, the "Wild Man" of Tenerife', in *Medical Cultures of the Early Modern Spanish Empire*, ed. John Slater et al. (Farnham, Surrey: Ashgate, 2014), 107–28.

36 Walter Putnam, '"Please Don't Feed the Natives": Human Zoos, Colonial Desire and Bodies on Display', *French Literature Series*, 39 (2013), 55–68; Charles S. Maier, 'The Human Zoo', in *A World Connecting, 1870–1945*, ed. Emily S. Rosenberg (Cambridge, MA: Belknap Press of Harvard University, 2012), 153–95; Kimberley A. Hamlin, 'The "Case of a Bearded Woman": Hypertrichosis and the Construction of Gender in the Age of Darwin', *American Quarterly*, 63 (2011), 955–81.

37 Louis Adolphus Duhring, *Case of a Bearded Woman* (New York: Putnams, 1877).

38 Hamlin, 'The "Case of a Bearded Woman"', 958.

39 *Macbeth*, I. iii. 43–5; Hamlin, 'The "Case of a Bearded Woman"', 958–9.

40 Rebecca M. Herzig, 'The Woman Beneath the Hair: Treating Hypertrichosis, 1870–1930', *National Women's Studies Association Journal*, 12 (2000), 50–66, at 54.

41 Ernest L. McEwen, 'The Problem of Hypertrichosis', *Journal of Cutaneous Diseases Including Syphilis*, 35 (1917), 829–36, at 830.

42 Christine Hope, 'Caucasian Female Body Hair and American Culture', *Journal of American Culture*, 5 (1982), 93–9.

43 Labre, 'The Brazilian Wax', 115.

44 Jennifer L. Bercaw-Pratt et al., 'The Incidence, Attitudes and Practices of the Removal of Pubic Hair as a Body Modification', *Journal of Pediatric and Adolescent Gynecology*, 25 (2012), 12–14.

45 Debra Herbenick et al., 'Pubic Hair Removal among Women in the United States: Prevalence, Methods, and Characteristics', *The Journal of Sexual Medicine*, 7 (2010), 3322–30.

46 Sara Ramsey et al., 'Pubic Hair and Sexuality: A Review', *The Journal of Sexual Medicine*, 6 (2009), 2102–110.
47 See Heinz Tschachler et al. (eds), *The EmBodyment of American Culture* (Münster: Lit 2003); Gail Dines, *Pornland: How Porn Has Hijacked our Sexuality* (Boston, MA: Beacon Press, 2010).
48 Diane E. Levin and Jean Kilbourne, *So Sexy So Soon: The New Sexualized Childhood and What Parents Can Do to Protect their Kids* (New York: Ballantine, 2008); M. Gigi Durham, *The Lolita Effect: The Media Sexualization of Young Girls and What we Can Do about it* (Woodstock, NY: Overlook Press, 2008); Sharna Olfman, *The Sexualization of Childhood* (Westport, CT: Praeger, 2009).
49 Roger Friedland, 'Looking through the Bushes: The Disappearance of Pubic Hair', *Huffington Post*, 13 June 2011, http://www.huffingtonpost.com/roger-friedland/women-pubic-hair_b_875465.html (accessed 1 July 2014).
50 Germaine Greer, *The Female Eunuch* (London: MacGibbon and Kee, 1970), 38, and Merran Toerien and Sue Wilkinson, 'Gender And Body Hair: Constructing The Feminine Woman', *Women's Studies International Forum*, 26 (2003), 333–44.
51 Friedland, 'Looking through the Bushes'.
52 Giulia Sissa, *Greek Virginity*, trans. Arthur Goldhammer (Cambridge, MA: Harvard University Press, 1990).
53 Kathleen Coyne Kelly, *Performing Virginity and Testing Chastity in the Middle Ages* (London: Routledge, 2000), 10–11.
54 Marina Warner, *Alone of All her Sex: The Myth and the Cult of the Virgin Mary* (Oxford: Oxford University Press, 2013); Miri Rubin, *Mother of God: A History of the Virgin Mary* (New York: Allen Lane, 2009).
55 Shirley Nielsen Blum, 'Hans Memling's Annunciation with Angelic Attendants', *Metropolitan Museum Journal*, 27 (1992), 43–58.
56 Montague R. James, trans., *The Apocryphal New Testament: Being the Apocryphal Gospels, Acts, Epistles and Apocalypses, with Other Narratives and Fragments* (Oxford: Clarendon Press, 1924), chs. 19–20.
57 Christa Grössinger, *Picturing Women in Late Medieval and Early Renaissance Art* (Manchester: Manchester University Press, 1997), 25.
58 See Marie H. Loughlin, *Hymeneutics: Interpreting Virginity on the Early Modern Stage* (Lewisburg: Bucknell University Press, 1997), 29.
59 Sara Read, *Menstruation and the Female Body in Early-Modern England* (Basingstoke: Palgrave Macmillan, 2013), 138–9.
60 Elizabeth A. Foyster, *Manhood in Early Modern England: Honour, Sex and Marriage* (London: Longman, 1999), 48.
61 Laura M. Carpenter, 'Like a Virgin ... Again?: Secondary Virginity as an Ongoing Gendered Social Construction', *Sexuality & Culture*, 15 (2011), 115–40, and Amani M. Awwad, 'Virginity Control and Gender-Based Violence in Turkey: Social

Constructionism of Patriarchy, Masculinity, and Sexual Purity', *International Journal of Humanities and Social Science*, I (2011), 105–10.

62 Dilek Cindoglu, 'Virginity Tests and Artificial Virginity in Modern Turkish Medicine', *Women's Studies International Forum*, 20 (1997), 253–61.

63 See http://www.hymenshop.com (accessed 1 July 2015).

64 Rebecca J. Cook and Bernard M. Dickens, 'Hymen Reconstruction: Ethical and Legal Issues', *International Journal of Gynecology & Obstetrics*, 107 (2009), 266–9.

65 Monica Christianson and Carola Eriksson, 'A Girl Thing: Perceptions concerning the Word "Hymen" among Young Swedish Women and Men', *Journal of Midwifery and Women's Health*, 56 (2011), 167–72.

66 Coyne Kelly, *Performing Virginity*, 26; Michel Thiery, 'Female Genital Organs in Vesalius' Iconography', in *Art of Vesalius*, ed. Robrecht van Hee (Antwerp: Garant, 2014), 161–80.

67 Coyne Kelly, *Performing Virginity*, 27.

68 Helkiah Crooke, *Mikrokosmographia: A Description of the Body of Man* (London: William Jaggard, 1615).

69 Margaret Ferguson, 'Hymeneal Instruction', in *Masculinities, Childhood, Violence: Attending to Early Modern Women*, ed. Amy E. Leonard and Karen L. Nelson (Newark: University of Delaware, 2011), 97–130.

70 Cited in Loughlin, *Hymeneutics*, 30.

71 Londa L. Schiebinger, *Nature's Body: Sexual Politics and the Making of Modern Science* (London: Pandora, 1993), 97.

72 Jacques Roger, *Buffon: A Life in Natural History*, trans. Sarah Lucille Bonnefoi, ed. Williams, L. Pearce (Ithaca, NY: Cornell University Press, 1997), 168.

73 Schiebinger, *Nature's Body*, 94.

74 Deanna Holtzman and Nancy Kulish, 'Nevermore: The Hymen and the Loss of Virginity', *Journal of the American Psychoanalytic Association*, 44 (1996), 303–32.

75 Naomi Wolf, *Vagina: A New Biography* (London: Virago, 2013).

76 Edward Erwin, *The Freud Encyclopaedia: Theory, Therapy and Culture* (London: Routledge, 2002), 593.

77 Sigmund Freud, 'Fetishism', in *On Sexuality: Three Essays on the Theory of Sexuality and Other Works*, trans. James Strachey et al., ed. Angela Richards (London: Penguin-Pelican, 1977; repr. Penguin, 1991), 354. [Ger. orig., *Drei Abhandlungen zur Sexualtheorie* (Leipzig and Vienna: Franck Deuticke, 1905).]

78 Sigmund Freud, 'Aledusa's Head', in *The Standard Edition of the Complete Psychological Works of Sigmund Freud*, ed. James Strachey, 24 vols (London: Hogarth, 1964), Vol. 18, 273–4.

79 Verrier Elwin, 'The Vagina Dentata Legend', *British Journal of Medical Psychology*, 19 (1943), 439–53.

80 Frank Samuel Caprio, *The Sexually Adequate Female* (New York: Citadel, 1953), 64.

81 Anne Koedt, *The Myth of the Vaginal Orgasm* (London: Women's Liberation Movement, 1968).

82 Natalie Angier, *Woman: An Intimate Biography* (Boston, MA: Houghton Mifflin, 1999).
83 Pedro Acién, 'Embryological Observations on the Female Genital Tract', *Human Reproduction*, 7 (1992), 437–45.
84 Mark D. Stringer and Innes Becker, 'Colombo and the Clitoris', *European Journal of Obstetrics and Gynecology and Reproductive Biology*, 151 (2010), 130–3.
85 Cited in Stringer and Becker, 'Colombo and the Clitoris', 131.
86 'Classic Pages in Obstetrics and Gynecology', *American Journal of Obstetric Gynecology*, 117 (1973), 144; Helen E. O'Connell et al., 'Anatomy of the Clitoris', *Journal of Urology*, 174 (2005), 1189–95, at 1193.
87 Andreas Vesalius, *Anatomicarum gabrielis falloppi observationum examen* (Venice: Francesco de' Franceschi da Siena, 1564), 143.
88 Katherine Park, 'The Rediscovery of the Clitoris', in *The Body in Parts: Fantasies of Corporeality in Early Modern Europe*, ed. David A. Hillman and Carla Mazzio (Routledge: New York, 1997), 170–93; Danielle Jacquart and Claude Thomasset, *Sexuality and Medicine in the Middle Ages* (Cambridge: Polity, 1988), 46.
89 Cited in O'Connell et al., 'Anatomy of the Clitoris', 1192.
90 O'Connell et al., 'Anatomy of the Clitoris', 1192.
91 Discussed in O'Connell et al., 'Anatomy of the Clitoris', 1193.
92 Georg Ludwig Kobelt, *Die männlichen und weibleichn Wollustorgane des Menschen und einiger Saügethiere* (Freiburg, 1844).
93 Cited in O'Connell et al., 'Anatomy of the Clitoris', 1193.
94 Stringer and Becker, 'Colombo and the Clitoris', 132.
95 See Susan Ekbery Stiritz, 'Cultural Cliteracy: Exposing the Contexts of Women's Not Coming', *Berkeley Journal of Gender, Law and Justice*, 23 (2008), 243–66.
96 Vincent Di Marino and Hubert Lepidi (eds), *Anatomic Study of the Clitoris and the Bulbo-Clitoral Organ* (New York: Springer, 2014), xii.
97 Helen E. O'Connell et al., 'Anatomical Relationship Between Urethra and Clitoris', *Journal of Urology*, 159 (1998), 1892–7.
98 Cited in O'Connell et al., 'Anatomy of the Clitoris', 1193.
99 Nancy Mann Kulish, 'The Mental Representation of the Clitoris: The Fear of Female Sexuality', *Psychoanalytic Inquiry*, 11 (1991), 511–36.
100 Deborah Gorham, *The Victorian Girl and the Feminine Ideal* (Hoboken, NJ: Taylor and Francis, 2012), ch. five.

Chapter 4. 'Soft and Tender' or 'Weighed down by Grief': The Emotional Heart

1 *OED*, 'Heart, n., int., and adv.'.
2 http://wellcomecollection.org./ (accessed 27 November 2015).
3 http://news.bbc.co.uk/1/hi/health/6977399.stm (accessed 4 September 2007).

4 Bound Alberti, *Matters of the Heart*.
5 On the impact of music on the heart, see Linda L. Chlan, 'Psychophysiologic Responses of Mechanically Ventilated Patients to Music: A Pilot Study', *American Journal of Critical Care*, 4 (1995), 233–8.
6 Bound Alberti, *Matters of the Heart*, introduction; Ashlee Neser, 'The Modern Heart and Narratives of Transplant', *English Studies in Africa*, 55 (2012), 50–63.
7 Ali Mazrui, 'The Poetics of a Transplanted Heart', *Transition*, 35 (1968), 51–9, at 56.
8 David O. Taylor et al., 'Registry of the International Society for Heart and Lung Transplantation: Twenty-Fourth Official Adult Heart Transplant Report 2007', *Journal of Heart and Lung Transplantation*, 26 (2007), 769–81.
9 Claire Sylvia with William Novak, *A Change of Heart: A Memoir* (London: Little, Brown, 1997); Paul Pearsall et al., 'Changes in Heart Transplant Recipients that Parallel the Personalities of their Donors', *Journal of Near-Death Studies*, 20 (2002), 191–206. Sadly, Claire passed away in 2009.
10 Benjamin Bunzel et al., 'Does Changing the Heart Mean Changing Personality? A Retrospective Inquiry on 47 Heart Transplant Patients', *Quality of Life Research*, 1 (1992), 251–6.
11 http://www.bhf.org.uk/heart-health/how-your-heart-works.aspx (accessed 12 October 2014).
12 Bound Alberti, *Matters of the Heart*, introduction.
13 See Geoffrey Gorham, 'Mind–Body Dualism and the Harvey–Descartes Controversy', *Journal of the History of Ideas*, 55 (1994), 211–34.
14 For a recent discussion see Brian Cummings and Freya Sierhuis, 'Introduction', in *Passions and Subjectivity in Early Modern Culture*, ed. Brian Cummings and Freya Sierhuis (Farnham: Ashgate, 2013), 1–12.
15 *Hamlet*, III. ii. 71.
16 Robert Burton, *The Anatomy Of Melancholy* (Oxford: John Lichfield and James Short for Henry Cripps, 1621), 152–3.
17 Pierre de la Primaudaye, *The French Académie*, trans. Thomas Bowes et al. (London: Adams, 1618), 471.
18 John Downame, *A Treatise of Anger* (London: William Welby, 1609), 3.
19 Thomas Wright, *Passions of the Minde*, 11, 8.
20 See the discussion in Lelland Joseph Rather, 'Old and New Views of the Emotions and Bodily Changes: Wright and Harvey versus Descartes, James and Cannon', *Clio Medica*, 1 (1965), 1–25, at 4. 4.
21 Normandi Ellis, *Awakening Osiris: A New Translation of the Egyptian Book of the Dead* (Grand Rapids, MI: Phanes, 1988).
22 Robert K. Ritner, 'The Cult of the Dead', in *Ancient Egypt*, ed. David P. Silverman (Oxford: Oxford University Press, 2003), 132–47, at 137.

23 Ritner, 'The Cult of the Dead', 138.
24 Ellis, *Awakening Osiris,* and T. J. Pettigrew, *A History of Egyptian Mummies* (London: Longman, Rees, Orme, Brown, Green and Longman, 1834), 52–3.
25 See Edwin Clarke and Charles Donald O'Malley, *The Human Brain and Spinal Cord: A Historical Study Illustrated by Writings from Antiquity to the Twentieth Century* (Berkeley, CA: University of California Press, 1968).
26 Walter Charleton, *Natural History of the Passions* (London: James Magnes, 1674), 70.
27 Downame, *A Treatise of Anger*, 56.
28 Charleton, *Natural History*, 151. On the 'animal spirits' see L. Stephen Jacyna, 'Animal Spirits and Eighteenth-Century British Medicine', in *The Comparison between Concepts of Life-Breath in East and West*, ed. Yosio Kawakita et al. (Tokyo: Ishiyaku EuroAmerica, 1995), 139–61.
29 Charleton, *Natural History*, 151.
30 See Charles Darwin, *The Expression of the Emotions in Man and Animals, with Photographic and Other Illustrations* (London: John Murray, 1872) and Paul Ekman (ed.), *Darwin and Facial Expression: A Century of Research in Review* (Cambridge, MA: Malor Books, 2006).
31 See Fay Bound Alberti, 'The Emotional Heart: Mind, Body and Soul', in *The Heart*, ed. James Peto (New Haven, CT: Yale University Press, 2006), 125–42.
32 See Saeed Changizi Ashtiyani and Mohsen Shamsi, 'The Discoverer of Pulmonary Blood Circulation: Ibn Nafis or William Harvey?' *Middle-East Journal of Scientific Research*, 18 (2013), 562–8.
33 Mohd Akmal et al., 'Ibn Nafis: A Forgotten Genius in the Discovery of Pulmonary Blood Circulation', *Heart Views*, 11 (2010), 26–30.
34 Stanley G. Schultz, 'William Harvey and the Circulation of the Blood: The Birth of a Scientific Revolution and Modern Physiology', *Physiology*, 17 (2002), 175–80, at 179.
35 William Harvey, *Exercitatio anatomica de motu cordis et sanguinis in animalibus* (Frankfurt am Main: Sumptibus Guilielmi Fitzeri, 1628).
36 Schultz, 'William Harvey and the Circulation of the Blood'.
37 René Descartes, *Discourse on Method and Meditations*, trans Elizabeth S. Haldane and G. R. T. Ross (Mineola, NY: Dover publications, 2003).
38 Moncef Berhouma, 'Beyond the Pineal Gland Assumption: A Neuroanatomical Appraisal of Dualism in Descartes' Philosophy', *Clinical Neurology and Neurosurgery*, 115 (2013), 1661–70.
39 William C. Aird, 'Discovery of the Cardiovascular System: From Galen to William Harvey', *Journal of Thrombosis and Haemostasis*, 9 (2011), 118–29.
40 Peter Harrison, 'Was There a Scientific Revolution?' *European Review*, 15 (2007), 445–57.

41 Daniel Garber, 'Philosophia, Historia, Mathematica: Shifting Sands in the Disciplinary Geography of the Seventeenth Century', in *Scientia In Early Modern Philosophy: Seventeenth-Century Thinkers On Demonstrative Knowledge From First Principles*, ed. Tom Sorell et al. (Dordrecht: Springer, 2010), 1–18, at 4.
42 On the construction of the metaphor of the heart as pump, and the links between scientific metaphor and social change more generally see Winifred Nöth (ed.), *Semiotics of the Media: State of the Art, Projects, and Perspectives* (Berlin: Mouton de Gruyter, 1997), 33.
43 Andrew Gregory, 'Harvey, Aristotle and the Weather Cycle', *Studies in History and Philosophy of Biological and Biomedical Sciences*, 32 (2001), 153–68.
44 See Otto Mayr, *Authority, Liberty and Automatic Machines in Early Modern Europe* (Baltimore, MD: Johns Hopkins University Press, 1986); Patricia S. Churchland, *Neurophysiology: Toward a Unified Science of the Mind/Brain* (Cambridge, MA: MIT Press, 1986); William Barrett, *Death of the Soul: From Descartes to the Computer* (New York: Doubleday, 1986).
45 Michel Foucault, *Language, Counter-Memory, Practice: Selected Essays and Interviews*, ed. Donald F. Bouchard (Ithaca, NY: Cornell University Press, 1977); Molly Andrews, 'Opening to the Original Contributions: Counter-Narratives and the Power to Oppose', in *Considering Counter-Narratives: Narrating, Resisting, Making Sense*, ed. Michael Bamberg and Molly Andrews (Amsterdam: Benjamins, 2004), 1–6.
46 Duncan Wu, *Romanticism: An Anthology* (Oxford: Wiley-Blackwell, 2012).
47 Arthur M. Z. Norman, 'Shelley's Heart', *Journal of the History of Medicine and Allied Sciences*, 10 (1955), 114.
48 Katherine Byrne, 'Consuming Flesh, Producing Fictions: Representations of Tuberculosis in Victorian Literature', PhD thesis, University of East Anglia, 2005.
49 Bound Alberti, *Matters of the Heart*, ch. 6.
50 See Kirstie Blair, *Victorian Poetry and the Culture of the Heart* (Oxford: Clarendon Press, 2006), 2.
51 Elizabeth Barrett Browning, *Sonnets from the Portuguese*, ed. William S. Paterson (London: Chapman and Hall, 1850; facs. edn of the BL MS, Barre, MA: Imprint Society, 1977); Alexander Mackenzie and Christina Rossetti, 'A Birthday' ('My Heart is like a Singing Bird'), in *Three Songs, The Poetry Written by Christina Rossetti; The Music Composed by A. C. Mackenzie*, 3 (London: Novello [1878] [CPM]); Harriet Martineau, *Deerbrook*, 3 vols (London: Moxon, 1839).
52 Martineau, *Deerbrook*, Vol. 2, 246.
53 Martineau, *Deerbrook*, Vol. 2, 124, 96.
54 Martineau, *Deerbrook*, Vol. 2, 137, 142.
55 Martineau, *Deerbrook*, Vol. 2, 96, 149.

56 For example, John Conolly et al. (eds), *The Cyclopaedia of Practical Medicine*, 4 vols (London: Sherwood, Gilbert and Piper, 1833–35).
57 Christopher Lawrence, 'Ancients and Moderns: The "New Cardiology" in Britain, 1880–1930', in *The Emergence of Modern Cardiology*, ed. William F. Bynum et al. (London: Wellcome Institute for the History of Medicine, 1985), 1–33.
58 Blair, *Victorian Poetry*, 27.
59 John F. Todaro et al., 'Prevalence of Anxiety Disorders in Men and Women with Established Coronary Heart Disease', *Journal of Cardiopulmonary Rehabilitation and Prevention*, 27 (2007), 86–91.
60 Martin Cowie, personal communication, 10 March 2010.
61 Francis Wells, personal communication, 8 March 2010.
62 H. Martyn et al., 'Medical Students' Responses to the Dissection of the Heart and Brain: A Dialogue on the Seat of the Soul', *Clinical Anatomy*, 25 (2012), 407–13.
63 Martyn et al., 'Medical Students' Responses', 411–12.
64 Martyn et al., 'Medical Students' Responses', 410.
65 http://www.heartmath.org (accessed 21 March 2015); Doc Lew Childre and Howard Martin with Donna Beech, *The HeartMath Solution: Proven Techniques for Developing Emotional Intelligence* (London: Piatkus, 1999).
66 J. Andrew Armour, 'Potential Clinical Relevance of the "Little Brain" on the Mammalian Heart', *Experimental Physiology*, 93 (2008), 165–76.
67 See N. Boyadjian, *The Heart: Its History, its Symbolism, its Iconography and its Diseases*, trans. Agnes Hall (Antwerp: Esco Books, 1985).

Chapter 5. Mind the Brain: From 'Cold Wet Matter' to the Motherboard

1 Francis Joseph Gall, *On the Functions of the Brain*, trans. Winslow Lewis (Boston, MA: Marsh, Capen & Lyon, 1835), 268.
2 John M. Harlow, *Recovery from the Passage of an Iron Bar through the Head* (Boston, MA: Clapp, 1869); Hanna Damasio et al., 'The Return of Phineas Gage: Clues about the Brain from the Skull of a Famous Patient', *Science*, 264 (1994), 1102–5.
3 Harlow, *Recovery*, 5–6.
4 John M. Harlow, 'Passage of an Iron Rod through the Head', *Boston Medical and Surgical Journal*, 39 (1848), 389–93.
5 Victoria Pitts-Taylor, 'Social Brains, Embodiment and Neuro-Interactionism', in the *Routledge Handbook of Body Studies*, ed. Bryan S. Turner (London: Routledge, 2014), 171–82.
6 Fernando Vidal, 'Person and Brain: A Historical Perspective from within the Christian Tradition', *Scripta Varia*, 109 (2007), 3–14, at 6.
7 Or, more recently, in the notion of neuroplasticity: Jeffrey M. Schwartz and Sharon Begley, *The Mind and the Brain: Neuroplasticity and the Power of Mental Force* (New York: Regan Books, 2002) and Brian Dolan, 'Soul Searching: A Brief History of the Mind/Body Debate in the Neurosciences', *Neurosurgical Focus*, 23 (2007), E2.

8 Warren D. TenHouten, *Emotion and Reason: Mind, Brain and the Social Domains of Work and Love* (Abingdon-on-Thames: Routledge, 2013), 95.
9 Cited in TenHouten, *Emotion and Reason*, 95.
10 *Othello*, IV. i. 271.
11 Arthur Lovejoy, *The Great Chain of Being: A Study of the History of an Idea* (Cambridge, MA: Harvard University Press, 1948); Frances Larson, *Severed: A History of Heads Lost and Heads Found* (London: Granta, 2014).
12 Harlow, *Recovery*, 18.
13 Harlow, *Recovery*, 6.
14 Harlow, *Recovery*, 6.
15 Harlow, *Recovery*, 8.
16 Harlow, *Recovery*, 9.
17 Harlow, *Recovery*, 10.
18 Harlow, *Recovery*, 13–14.
19 Paul Broca, 'Remarks on the Seat of the Faculty of Articulated Language, Following an Observation of Aphemia (Loss of Speech)', *Bulletin de la Société Anatomique*, 6 (1861), 330–57; Ennis A. Berker et al., 'Translation of Broca's 1865 Report: Localization of Speech in the Third Left Frontal Convolution', *Archives of Neurology*, 43 (1986), 1065–72; Carl Wernicke, *Der aphasische Symptomencomplex* (Breslau: Cohn und Weigert, 1874).
20 Harlow, *Recovery*, 10.
21 Henry Jacob Bigelow, 'Dr. Harlow's Case of Recovery from the Passage of an Iron Bar through the Head', *American Journal of the Medical Sciences*, 19 (1850), 13–22.
22 Bigelow, 'Dr Harlow's Case', 19.
23 David Ferrier, 'The Goulstonian Lectures on the Localisation of Cerebral Disease', *British Medical Journal*, 1 (1878), 591–5.
24 Kieran O'Driscoll and John Paul Leach, ' "No Longer Gage": An Iron Bar through the Head', *BMJ*, 317 (1998), 1673–4, at 1674.
25 Bryan Kolb and Ian Q. Whishaw, 'Brain Plasticity and Behavior', *Annual Review of Psychology*, 49 (1998), 43–64.
26 M. James C. Crabbe (ed.), *From Soul to Self* (London: Routledge, 1999).
27 Pavel Gregoric, *Aristotle on the Common Sense* (Oxford: University of Oxford, 2003).
28 Robert W. Doty, 'Alkmaion's Discovery that Brain Creates Mind: A Revolution in Human Knowledge Comparable to that of Copernicus and of Darwin', *Neuroscience*, 147 (2007), 561–8.
29 Christos Yapijakis, 'Hippocrates of Kos, the Father of Clinical Medicine, and Asclepiades of Bithynia, the Father of Molecular Medicine', *In Vivo*, 23 (2009), 507–14.
30 Hippocrates cited in Stanley Finger, *Origins of Neuroscience: A History of Explorations into Brain Function* (Oxford: Oxford University Press, 2001), 13.

On Hippocrates' influence on modern neurological anatomy, see Ioannis G. Panourias et al., 'The Ancient Hellenic and Hippocratic Origins of Head and Brain Terminology', *Clinical Anatomy*, 25 (2012), 548–58.

31 But see the discussion in Tomislav Breitenfeld et al., 'Hippocrates: The Forefather of Neurology', *Neurological Sciences*, 35 (2014), 1349–52.

32 Patrick Berche and Jean-Jacques Lefrere, 'Herophilus and Erasistratus: The First Exploration of the Human Body', *Presse médicale*, 40 (2011), 535–9.

33 F. Clifford Rose, 'Cerebral Localization in Antiquity', *Journal of the History of the Neurosciences*, 18 (2009), 239–47; J. M. S. Pearce, 'The Neurology of Erasistratus', *Journal of Neurological Disorders*, 1 (2013), 1–3; Noel Si-Yang Bay and Boon-Huat Bay, 'Greek Anatomist Herophilus: The Father of Anatomy', *Anatomy & Cell Biology*, 43 (2010), 280–3.

34 Andrew P. Wickens, *A History of the Brain: From Stone Age Surgery to Modern Neuroscience* (London: Routledge, 2014), 33.

35 Tullio Manzoni, 'The Cerebral Ventricles, the Animal Spirits and the Dawn of Brain Localization of Function', *Archives italiennes de biologie*, 136 (1998), 103–52.

36 See Maud W. Gleason, 'Shock and Awe: The Performance Dimension of Galen's Anatomy Demonstrations', in *Galen and the World of Knowledge*, ed. Christopher Gill et al. (Cambridge and New York: Cambridge University Press, 2009), 85–114.

37 Jonathan Sawday, *The Body Emblazoned: Dissection and the Human Body in Renaissance Culture* (London: Routledge, 2013); S. Ryan Gregory and Thomas R. Cole, 'The Changing Role of Dissection in Medical Education', *Journal of the American Medical Association*, 287 (2002), 1180–1.

38 Cited in Carlos Guillermo de Gutiérrez-Mahoney and Mannie M. Schechter, 'The Myth of the *Rete Mirabile* in Man', *Neuroradiology*, 4 (1972), 141–58; Jacopo Berengario da Carpi, *A Short Introduction to Anatomy (Isagogae Breves)*, trans. L. R. Lind, ed. Paul G. Roofe (Chicago, IL: University of Chicago Press, 1959; repr. New York: Kraus 1969); Richard Sugg, *The Smoke of the Soul: Medicine, Physiology and Religion in Early Modern England* (Basingstoke: Palgrave, Macmillan 2013), 3–4.

39 Jacopo Berengario da Carpi, *Anatomia carpi isagoge breves perlucide ac uberime, in anatomiam humani corporis* (Venice: de Vitalibus, 1535).

40 Benoit Bataille, et al., 'The Significance of the Rete Mirabile in Vesalius's Work: An Example of the Dangers of Inductive Inference in Medicine', *Neurosurgery*, 60 (2007), 761–8.

41 Jose Van Laere, '[Vesalius and the Nervous System]', *Verhandelingen-Koninklijke academie voor geneeskunde van Belgie*, 55 (1992), 533–76.

42 René Descartes, *The Passions of the Soul* (London: J. Martin and J. Ridley, 1650).

43 Ku-ming Kevin Chang, 'Alchemy as Studies of Life and Matter: Reconsidering the Place of Vitalism in Early Modern Chymistry', *Isis*, 102 (2011), 322–9; Catherine Packham, *Eighteenth-Century Vitalism: Bodies, Culture, Politics* (Basingstoke: Palgrave Macmillan, 2012).

44 James P. B. O'Connor, 'Thomas Willis and the Background to *Cerebri Anatome*', *Journal of the Royal Society of Medicine*, 96 (2003), 139–43.

45 Guilielmi Harvei (William Harvey), *Exercitatio anatomica de motu cordis et sanguinis in animalibus* (Frankfurt am Main: Sumptibus Guilielmi Fitzeri, 1628).

46 Christopher Hill, 'William Harvey and the Idea of Monarchy', *Past and Present*, 27 (1964), 54–72.

47 O'Connor, 'Thomas Willis', 141–3.

48 Hubert Steinke, *Irritating Experiments: Haller's Concept and the European Controversy on Irritability and Sensibility, 1750–90* (Amsterdam: Rodopi, 2005).

49 Albrecht von Haller, *A Dissertation on the Sensible and Irritable Parts of Animals* (London: Nourse, 1755).

50 C. U. M. Smith, 'Understanding the Nervous System in the 18th Century', *Handbook of Clinical Neurology*, 95 (2009), 107–14.

51 Robert Whytt, *Physiological Essays* (Edinburgh: Hamilton, Balfour and Neill, 1761); Eugenio Frixione, 'Irritable Glue: The Haller–Whytt Controversy on the Mechanism of Muscle Contraction', in *Brain, Mind and Medicine: Essays in Eighteenth-Century Neuroscience*, ed. Harry Whitaker et al. (New York: Springer US, 2007), 115–24.

52 Wickens, *History of the Brain*.

53 Carolyn Thomas de la Peña, *The Body Electric: How Strange Machines Built the Modern American* (New York: New York University Press, 2003).

54 Alvaro Pascual-Leone and Timothy Wagner, 'A Brief Summary of the History of Noninvasive Brain Stimulation', *Annual Review of Medical Engineering*, 9 (2007), 527–65.

55 Luigi Galvani, *Commentary on the Effects of Electricity on Muscular Motion*, trans. Margaret Glover Foley (Norwalk, CT: Burndy Library, 1953; repr., 1954). [Latin orig., *De viribus electricitatis in motu musculari commentarius* (Bologna: Instituti Scientiarum, 1791).]

56 Marco Piccolino and Nicholas J. Wade. 'The Frog's Dancing Master: Science, Séances, and the Transmission of Myths', *Journal of the History of the Neurosciences*, 22 (2013), 79–95.

57 Julian North, 'Shelley Revitalized: Biography and the Reanimated Body', *European Romantic Review*, 21 (2010), 751–70; Mary Shelley, *Frankenstein, or, The Modern Prometheus: The 1818 text*, ed. Marilyn Butler (London: William Pickering, 1993); Frederic L. Holmes, 'The Old Martyr of Science: The Frog in Experimental Physiology', *Journal of the History of Biology*, 26 (1993), 311–28.

58 Iwan Rhys Morus, '"The Nervous System of Britain": Space, Time and the Electric Telegraph in the Victorian Age', *The British Journal for the History of Science*, 33 (2000), 455–75.

59 Edwin Clarke and L. Stephen Jacyna, *Nineteenth-Century Origins of Neuroscientific Concepts* (Berkeley, CA, and London: University of California Press, 1987), 81.

60 Charles Bell, *Idea of a New Anatomy of the Brain: Submitted for the Observations of his Friends* (London: Strahan and Preston, 1811); François Magendie, 'Expériences sur les fonctions des racines des nerfs qui naissent de la moelle épinière', *Journal de Physiologie Expérimentale et Pathologie*, 2 (1822), 366–71; Alexander Walker, *Documents and Dates of Modern Discoveries in the Nervous System* (London: J. Churchill, 1839; repr. Metuchen, NJ: Scarecrow Reprints, 1973).
61 William James, 'What is an Emotion?' *Mind*, 9 (1884), 188–205, at 188.
62 James, 'Emotion', 189–90.
63 Walter B. Cannon, 'The James-Lange Theory of Emotions: A Critical Examination and an Alternative Theory', *American Journal of Psychology*, 39 (1927), 106–34.
64 Otniel E. Dror, 'The Affect of Experiment: The Turn to Emotions in Anglo-American Physiology, 1900–1940', *Isis* (1999), 205–37, at 218; Elin L. Wolfe, *Walter B. Cannon: Science and Society* (Boston MA: Harvard University Press, 2000).
65 Finger, *Origins of Neuroscience*, 271.
66 On the complex relationship between Darwin and Freud, see Lucille B. Ritvo, *Darwin's Influence on Freud: A Tale of Two Cities* (New Haven, CT: Yale University Press, 1990).
67 Finger, *Origins of Neuroscience*, 271–3.
68 Francis Hutcheson, *On the Nature and Conduct of the Passions with Illustrations on the Moral Sense*, ed. Andrew Ward (1728; Manchester: Clinamen, 1999).
69 Franz Joseph Gall, 'On Phrenology, the Localization of the Functions of the Brain', in *A Source Book in the History of Psychology*, ed. R. J. Herrnstein and E. G. Boring (Cambridge, MA: Harvard University Press, 1965), 211–20. See also D. Hothersall, *History of Psychology* (New York: McGraw-Hill, 2004), 89–90.
70 Francis Joseph Gall, *On the Functions of the Brain* (Boston, MA: Marsh, Capen and Lyon, 1835), 268.
71 Gall, *On the Functions of the Brain*, 29–30; 112; Anthelme Richerand, *The Elements of Physiology*, trans. Robert Kerrison (London: John Murray, 1803); Jennifer Radden, 'Lumps and Bumps: Kantian Faculty Psychology, Phrenology, and Twentieth-Century Psychiatric Classification', *Philosophy, Psychiatry and Psychology* 3 (1996), 1–14; Charles G. Gross, *Brain, Vision, Memory: Tales in the History of Neuroscience* (Cambridge, MA; MIT Press, 1998), 54.
72 Cesar Lombroso, *Criminal Man according to the Classification of Cesare Lombroso* (New York: Putnam, 1911); Nikolaas N. Oosterhof, 'The Functional Basis of Face Evaluation', *Proceedings of the National Academy of Sciences*, 105 (2008), 11087–92.
73 Georges John Romanes, 'Mental Differences between Men and Women', *Nineteenth Century*, 21 (1887), 654–72; Cynthia Eagle Russett, *Sexual Science: The Victorian Construction of Womanhood* (Cambridge, MA: Harvard University Press, 1991), 35–6.
74 Rachel Malane, *Sex in Mind: The Gendered Brain in Nineteenth-Century Literature and Mental Sciences* (New York: Peter Lang, 2005), 5.

75 Russett, *Sexual Science*, 19–21.
76 Bernard Hollander, *The Revival of Phrenology: The Mental Functions of the Brain* (London: Richards, 1901); Richard Twine, 'Physiognomy, Phrenology and the Temporality of the Body', *Body & Society*, 8 (2002), 67–88; Ann Fabian, *The Skull Collectors: Race, Science and America's Unburied Dead* (Chicago, IL: University of Chicago Press, 2010).
77 Cited in Stephen Jay Gould, 'Women's Brains', in *The Panda's Thumb: More Reflections in Natural History* (Harmondsworth: Penguin, 1983), 152–9, at 158.
78 Fabian, *The Skull Collectors*.
79 Sandra Blakeslee, 'A Small Part of the Brain and its Profound Effects', *New York Times*, 6 February 2007; Amber N. V. Ruigrok et al., 'A Meta-Analysis of Sex Differences in Human Brain Structure', *Neuroscience & Biobehavioral Reviews*, 39 (2014), 34–50.
80 Louann Brizendine, *The Female Brain* (London: Bantam, 2006), 155.
81 Barbara Annis and John Gray, *Work with Me: The 8 Blind Spots Between Men and Women in Business* (Basingstoke: Palgrave MacMillan, 2013); Craig Howard Kinsley and Kelly G. Lambert, 'The Maternal Brain', *Scientific American*, 294 (2006), 72–9.
82 See Cordelia Fine et al., 'Plasticity, Plasticity, Plasticity . . . and the Rigid Problem of Sex', *Trends in Cognitive Sciences*, 17 (2013), 550–1.
83 Edward S. Reed, *From Soul to Mind: The Emergence of Psychology from Erasmus Darwin to William James* (New Haven, CT: Yale University Press, 1997).
84 Owen Flanagan, *The Problem of the Soul: Two Visions of Mind and How to Reconcile Them* (New York: Basic Books, 2002), xii.
85 Thomas Dixon, *From Passions to Emotions: The Creation of a Secular Psychological Category* (Cambridge: Cambridge University Press, 2003).
86 *The Brain That Wouldn't Die*, dir. Joseph Green, (USA: Rex Carlton, 1962).
87 In February 2015 an Italian team claimed that the first head transplant was only two years away. Broadsheets and tabloids alike pounced at the story with predictable interest: http://www.theguardian.com/society/2015/feb/25/first-full-body-transplant-two-years-away-surgeon-claim (accessed 1 March 2015).
88 Robert J. White et al, 'Cephalic Exchange Transplantation in the Monkey', *Surgery*, 70 (1971), 135–9.
89 Robert J. White et al., 'Primate Cephalic Transplantation: Neurogenic Separation, Vascular Association', *Transplantation Proceedings*, 3 (1971) 602–4.
90 Letter to the editor from Carla Bennett, Senior Writer, People for the Ethical Treatment of Animals (PETA), 21 August 1995.
91 'Frankenstein Fears After Head Transplant', BBC News, Friday, 6 April 2001: http://news.bbc.co.uk/1/hi/health/1263758.stm (accessed 31 October 2014).
92 Quoted in 'Frankenstein Fears after Head Transplant'.
93 See Bound Alberti, *Matters of the Heart*, introduction.

94 See John R. Searle, 'The Self as a Problem in Philosophy and Neurobiology', in *The Lost Self: Pathologies of the Brain and Identity*, ed. Todd E. Feinberg et al. (Oxford: Oxford University Press, 2005), ch. two.

95 Martyn et al., 'Medical Students' Responses', 407–13, at 409.

96 Martyn et al., 'Medical Students' Responses', 410.

97 Henry Marsh, *Do No Harm: Stories of Life, Death and Brain Surgery* (London: Weidenfeld & Nicolson, 2014), 1.

98 Sunil K. Pandya, 'Understanding Brain, Mind and Soul: Contributions from Neurology and Neurosurgery', *Mens Sana Monographs*, 9 (2011), 129–49.

99 Slawomir J. Nasuto et al., 'Communication as an Emergent Metaphor for Neuronal Operation', in *Computation for Metaphors, Analogy, and Agents*, ed. Chrystopher L. Nehaniv (Berlin and New York: Springer 1999), 365–79.

Chapter 6. From Excrement to Boundary: Touching on the Skin

1 Harper Lee, *To Kill a Mockingbird* (Bath: Chivers Press, 1992), 29.

2 William Lawrence and Henry Herbert Southey, 'Two Cases of the True Elephantiasis, or Lepra Arabum', *Medico-Chirurgical Transactions*, 6 (1815), 209–20, at 210.

3 John A. McGrath and Jouni Uitto, 'Anatomy and Organization of Human Skin', in *Rook's Textbook of Dermatology*, 8th edn, ed. Tony Burns et al. (Oxford: Wiley-Blackwell, 2010), Vol. I, 1–53.

4 Sarah Kay, 'Original Skin: Flaying, Reading, and Thinking in the Legend of Saint Bartholomew and Other Works', *Journal of Medieval and Early Modern Studies*, 36 (2006), 35–74.

5 Steven Connor, *The Book of Skin* (Reaktion: London, 2004), ch. 1.

6 See Nina G. Jablonski, *Skin: A Natural History* (London: University of California Press, 2006), 3.

7 Cited in Connor, *The Book of Skin*, 9.

8 David R. Bickers et al., 'The Burden of Skin Diseases, 2004: A Joint Project of the American Academy of Dermatology Association and the Society for Investigative Dermatology', *Journal of the American Academy of Dermatology*, 55 (2006), 490–500.

9 Thomas Bateman, *Delineations of Cutaneous Diseases* (London: Longman, 1817), plate LXVIII.

10 Lawrence and Southey, 'Two Cases of the True Elephantiasis', 212.

11 Lawrence and Southey, 'Two Cases of the True Elephantiasis', 217. On letter writing conventions, see Fay Bound, 'Writing the Self? Love and the Letter in England, *c.*1660–*c.*1760', *Literature and History*, 11 (2002), 1–19.

12 On the provision of charitable treatment at London hospitals, see Susan C. Lawrence, *Charitable Knowledge: Hospital Pupils and Practitioners in Eighteenth-Century London* (Cambridge: Cambridge University Press, 1996).

13 See Nora Crook and Derek Guiton, *Shelley's Venomed Melody* (Cambridge: Cambridge University Press, 1986), 94.

14 Avicenna, *The Canon of Medicine* (New York: AMS Press, 1973), 445.
15 Cited by Connor, *The Book of Skin*, 12.
16 And see A. A. Diamandopoulos et al., 'The Human Skin: A Meeting Ground for the Ideas about Macrocosm and Microcosm in Ancient and Medieval Greek literature', *Vesalius*, 7 (2001), 94–101.
17 On ancient medicine and the non-naturals see Vivian Nutton, *Ancient Medicine* (London and New York: Routledge, 2013), 306.
18 On the rise in dermatological problems related to employment see James Cumming, 'Remarks on Psoriasis', *British Medical Journal*, 2 (1878), 353.
19 On Linnaeus, see William Thomas Stearn, *An Introduction to the Species Plantarum and Cognate Botanical Works of Carl Linnaeus* (London: Ray Society, 1957). On Willan's work at the Carey Street Dispensary see Andrzej Grzybowski and Lawrence Charles Parish, 'Robert Willan: Pioneer in Morphology', *Clinical Dermatology*, 29 (2011), 125–9.
20 Thomas Bateman, *A Practical Synopsis of Cutaneous Diseases: According to the Arrangement of Dr Willan* (London: Longman, Hurst, Rees, Orme and Brown, 1812).
21 See Kim F. Hall, '"These Bastard Signs of Fair": Literary Whiteness in Shakespeare's Sonnets', in *Post-Colonial Shakespeares*, ed. Ania Loomba and Martin Orkin (London and New York: Routledge, 1998), 64–83.
22 Kimberley Poitevin, 'Inventing Whiteness: Cosmetics, Race and Women in Early Modern England', *Journal for Early Modern Cultural Studies*, 11 (2011), 58–89.
23 Discussed in Nicholas Hudson, 'From "Nation" to "Race": The Origin of Racial Classification in Eighteenth-Century Thought', *Eighteenth-Century Studies*, 29 (1996), 247–64.
24 Anthony Gerard Barthelemy, *Black Face, Maligned Race: The Representation of Blacks in English Drama from Shakespeare to Southerne* (Baton Rouge, LA: Louisiana State University Press, 1987), 2–3.
25 Philip C. Kolin, 'Blackness Made Visible: A Survey of *Othello* in Criticism, on Stage, and on Screen', in *Othello: New Critical Essays*, ed. Philip C. Kolin (Abingdon-on-Thames: Routledge, 2002), 1–88.
26 Cited in Barthelemy, *Black Face, Maligned Race*, 5.
27 See Robert Bernasconi and Tommy Lott (eds), *The Idea of Race* (Indianapolis: Hackett, 2000).
28 Carolus Linnaeus, *Systema Naturae*, 2nd edn (Stockholm: Kiesewetter, 1740), and see Roxann Wheeler, *The Complexion of Race: Categories of Difference in Eighteenth-Century British Culture* (Philadelphia, PA: University of Pennsylvania Press, 2000), 22–4.
29 Johann Friedrich Blumenbach, *On the Natural Varieties of Mankind: De generis humani varietate nativa*, trans. and ed. Thomas Bendyshe (New York: Bergman Publishers, 1969).
30 Hudson, 'From "Nation" to "Race"', 252.

31 See Peter Hanns Reill, 'The Legacy of the Scientific Revolution: Science and the Enlightenment', in *The Cambridge History of Science*, iv: *Eighteenth-Century Science*, ed. Roy Porter (Cambridge: Cambridge University Press, 2003), 23–43.
32 See Paul Lawrence Farber, *Mixing Races: From Scientific Racism to Modern Evolution Ideas* (Baltimore, MD: Johns Hopkins University Press, 2011).
33 Cited in Tania Das Gupta et al. (eds), *Race and Racialization: Essential Readings* (Toronto: Canadian Scholars' Press, 2007), 25.
34 Portraiture as a whole focused intently on the colour and texture of the skin as canvas of identity. See Mechthild Fend, 'Bodily and Pictorial Surfaces: Skin in French Art and Medicine, 1790–1860', *Art History*, 28 (2005), 311–39.
35 Eric P. H. Li et al., 'Skin Lightening and Beauty in Four Asian Cultures', *Advances in Consumer Research*, 35 (2008), 444–9, and Christopher A. D. Charles, 'Skin Bleaching, Self-hate, and Black Identity in Jamaica', *Journal of Black Studies*, 33 (2003), 711–28; Pascal Del Giudice and Yves Pinier, 'The Widespread Use of Skin Lightening Creams in Senegal: A Persistent Public Health Problem in West Africa', *International Journal of Dermatology*, 41 (2002), 69–72.
36 Yaba Amgborale Blay, 'Skin Bleaching and Global White Supremacy: By Way of Introduction', *Journal of Pan African Studies*, 4 (2011), 4–46.
37 Advertisement for Pear's Soap, 1899, discussed in Blay, 'Skin Bleaching', 17.
38 See Angela Rosenthal, 'Visceral Culture: Blushing and the Legibility of Whiteness in Eighteenth-Century British Portraiture', *Art History*, 27 (2004), 563–92.
39 Charles Darwin, *The Expression of the Emotions in Man and Animals*, ed. Paul Ekman (London: Fontana, 1999), 310.
40 Thomas H. Burgess, *The Physiology or Mechanism of Blushing: Illustrative of the Influence of Mental Emotion on the Capillary Circulation* (London: Churchill, 1839), iii.
41 Charles Bell, *Essays on the Anatomy of Expression in Painting* (London: John Murray, 1806), 96.
42 Burgess, *The Physiology or Mechanism of Blushing*, 26.
43 Darwin, *The Expression of the Emotions*, 321.
44 See Angela Simon and Stephanie Shields, 'Does Complexion Color Affect the Experience of Blushing?' *Journal of Social Behavior & Personality*, 11 (1996), 177–88, and Felix I. S. Konotey-Ahulu, 'Blushing in Black Skin', *Journal of Cosmetic Dermatology*, 2 (2003), 59–60.
45 See Cristiano Castelfranchi and Isabella Poggi, 'Blushing as a Discourse: Was Darwin Wrong?', in *Shyness and Embarrassment: Perspectives from Social Psychology*, ed. W. Roy Crozier (Cambridge: Cambridge University Press, 1990), 230–54.
46 Konotey-Ahulu, 'Blushing in Black Skin', 60.
47 Bell, *Essays on the Anatomy of Expression*, 95.
48 On the nineteenth-century science of emotion, see Otniel Dror, 'The Scientific Image of Emotion: Experience and Technologies of Inscription', *Configurations*, 7 (1999), 355–401.

49 Fay Bound Alberti, 'Emotion Theory and Medical History', in *Medicine, Emotion and Disease, 1700–1950*, ed. Fay Bound Alberti (Palgrave Macmillan: Basingstoke, 2006), xiii–xxviii.

50 See Otniel Dror, 'The Affect of Experiment: The Turn to Emotions in Anglo-American Physiology, 1900–1940', *Isis*, 90 (1999), 205–37.

51 The limits of GSR were identified as early as the 1950s, though it is still used today. See Robert A. McCleary, 'The Nature of the Galvanic Skin Response', *Psychological Bulletin*, 47 (1950), 97–117.

52 Courtenay Young, 'The Science of Body Psychotherapy: The Science of Body Psychotherapy Today, Part 1: A Background History', *The USA Body Psychotherapy Journal Editorial*, 8 (2009), 5–15.

53 See Frederick Peterson and Carl Gustav Jung, 'Psycho-Physical Investigations with the Galvanometer and Pneumograph in Normal and Insane Individuals', *Brain*, 30 (1907), 153–218.

54 See Geoffrey C. Bunn, *The Truth Machine: A Social History of the Lie Detector* (Baltimore, MD: Johns Hopkins University Press, 2012) and Hedwig Lewis, *Body Language: A Guide for Professionals*, 3rd edn (New Delhi: Sage, 2012).

55 Robert E. Rankin and Donald T. Campbell, 'Galvanic Skin Response to Negro and White Experimenters', *The Journal of Abnormal and Social Psychology*, 51 (1955), 30–3.

56 Leslie E. Fisher and Harry Kotses, 'Race Differences and Experimenter Race Effect in Galvanic Skin Response', *Psychophysiology*, 10 (1973), 578–82, abstract.

57 See Leslie A. Zebrowitz et al. 'Facial Resemblance to Emotions: Group Differences, Impression Effects, and Race Stereotypes', *Journal of Personality and Social Psychology*, 98 (2010), 175–89.

58 Robert L. Rubinstein and Sarah Canham, 'Aging Skin in Sociocultural Perspective', in *Skin Aging Handbook: An Integrated Approach to Biochemistry and Product Development*, ed. Nava Dayan (Norwich, NY: William Andrew, 2008), 3–15, at 8.

59 Enzo Berardesca et al., 'In Vivo Biophysical Characterization of Skin Physiological Differences in Races', *Dermatologica*, 182 (1991), 89–93, abstract.

60 Naissan O. Wesley and Howard I. Maibach, 'Racial (Ethnic) Differences in Skin Properties: The Objective Data', *American Journal of Clinical Dermatology*, 4 (2003), 843–60.

61 For an introduction see Jan Caplan (ed.), *Written on the Body: The Tattoo in European and American History* (London: Reaktion, 2000).

62 See John Koo and Andrew Lebwohl, 'Psycho Dermatology: The Mind and Skin Connection', *American Family Physician*, 64 (2001), 1873–8.

63 On fears of ageing in the West, see the introduction in Abigail T. Brooks, 'Aesthetic Anti-Ageing Surgery and Technology: Women's Friend or Foe?' *Sociology of Health and Illness*, 32 (2010), 238–57.

64 See Pamela H. Mitchell et al., 'Critically Ill Children: The Importance of Touch in a High-Technology Environment', *Nursing Administration Quarterly*, 9 (1985), 38–46.

65 See Diana Adis Tahhan, 'Blurring the Boundaries Between Bodies: Skinship and Bodily Intimacy in Japan', *Japanese Studies*, 30 (2010), 215–30.

66 Matthew J. Hertenstein et al., 'The Communication of Emotion via Touch', *Emotion*, 9 (2009), 566–73.

Chapter 7. Tongue-Tied? From Nagging Wives to a Question of Taste

1 *Hamlet*, V. i. 75–9.

2 *Hamlet*, V. i. 18–186.

3 Ailsa Grant Ferguson, '"Tis now the very witching time of night": Halloween Horror and the *Memento Mori* in Hamlet (2000)', *Journal of Adaptation in Film & Performance*, 5 (2012), 127–47.

4 See Carla Mazzio, 'Sins of the Tongue in Early Modern England', *Modern Language Studies*, 28 (1998), 95–124, at 95.

5 John Heath, *The Talking Greeks: Speech, Animals and the Other in Homer, Aeschylus and Plato* (Cambridge and New York: Cambridge University Press, 2005), 8.

6 Cited in Stephen Greenblatt, *Learning to Curse: Essays in Early Modern Culture* (New York: Routledge, 1992), 23.

7 Keith Gunderson, 'Descartes, La Mettrie, Language, And Machines', *Philosophy*, 39 (1964), 193–222.

8 e.g. Susan Bennett, *Performing Nostalgia: Shifting Shakespeare and the Contemporary Past* (London: Routledge, 1996), ch. 4.

9 Ernst Cassirer, *Language and Myth*, trans. Susanne K. Langer (New York: Harper, 1946), 45–6.

10 *Hamlet*, II. ii. 304–7.

11 Thomas Adams, 'The Management of the Tongue', in *The Works of Thomas Adams; Being the Sum of his Sermons, Meditations and other Divine and Moral Discourses; with Memoir by Joseph Angus* (Edinburgh: Nichol, 1861–2), 10, 12–13.

12 For instance George Webbe, Bishop of Limerick, *The Arraignement of an Unruly Tongue* (London, 1619), the Puritan clergyman William Perkins, *A Direction for the Government of the Tongue According to God's Word* (Cambridge, 1593), and the English nonconformist Edward Reyner, *Rules for the Government of the Tongue* (London, 1656).

13 1 Cor.: 14.2.

14 Koran: 3.78.

15 Jerry Saltz, 'Art at Arm's Length: A History of the Selfie', *New York Magazine*, 47 (2014), 71–5.

16 For an introduction see Weston LaBarre, 'The Cultural Basis of Emotions and Gestures', *Journal of Personality*, 16 (1947), 49–68.

17 Nathalie Vienne-Guerrin (ed.), *The Unruly Tongue in Early Modern England: Three Treatises* (Madison, NJ: Fairleigh Dickinson University Press, 2012), xxii.

18 Jean de Marconville, *A Treatise of the Good and Evell Tounge*, in Vienne-Guerrin, *The Unruly Tongue*, xxxii.

19 Marconville, *Treatise*, xxxiv.
20 Vienne-Guerrin, *The Unruly Tongue*, xxxiv.
21 Vienne-Guerrin, *The Unruly Tongue* xxxii.
22 Helkiah Crooke, *Mikrokosmographia: A Description of the Body of Man. Together with the Controversies thereto Belonging, Collected and Translated out of all the Best Authors of Anatomy* (London: William Jaggard, 1615).
23 Crooke, *Mikrokosmographia*, 328.
24 Crooke, *Mikrokosmographia*, 328.
25 Pierre Louis Verdier, *An Abstract of the Human Body*, trans, Dale Ingram (London: John Clarke, 1753), 28.
26 Verdier, *Abstract of the Human Body*, 183.
27 See Lindsay Kaplan, *The Culture of Slander in Early Modern England* (Cambridge: Cambridge University Press, 1997), 2; and Fay Bound [Alberti], '"An Angry and Malicious Mind?" Narratives of Slander at the Church Courts of York, *c.*1660–*c.*1760', *History Workshop Journal*, 56 (2003), 59–77.
28 Laura Gowing, 'Gender and the Language of Insult in Early Modern London', *History Workshop Journal*, 35 (1993), 1–21, 2; Laura Gowing, *Domestic Dangers: Women, Words and Sex in Early Modern London* (Oxford, 1996).
29 Discussed in Gowing, 'Gender and the Language of Insult', 8.
30 See Elizabeth A. Foyster, *Manhood in Early Modern England: Honour, Sex and Marriage* (London: Longman, 1999; Routledge, 2014), 7–8.
31 Paul Lorrain, *The Ordinary of Newgate, his Account of the Behaviour, Confessions, and Dying Speeches of the Malefactors that were Executed at Tyburn, on Friday the 19th July, 1706* (London: Dryden Leach, 1706).
32 [Laurent Bordelon], *The Management of the Tongue* (London: D. Leach, 1706), title page.
33 Results of keyword search for 'tongue', www.oldbaileyonline.org (accessed 1 February 2015).
34 https://www.oldbaileyonline.org/browse.jsp?div=t16760510-1 (accessed 9 December 2015).
35 https://www.oldbaileyonline.org/browse.jsp?div=OA16790122 (accessed 9 December 2015).
36 https://www.oldbaileyonline.org/browse.jsp?div=t16940418-8 (accessed 9 December 2015).
37 https://www.oldbaileyonline.org/browse.jsp?div=t16940418-8 (accessed 9 December 2015).
38 'sharp Tongue': http://www.oldbaileyonline.org/browse.jsp?div=OA16850904 (accessed 9 December 2015); William Shakespeare, *Taming of the Shrew*, I. ii. 96.
39 See Lynda E. Boose, 'Scolding Brides and Bridling Scolds: Taming the Woman's Unruly Member', *Shakespeare Quarterly*, 42 (1991), 179–213.
40 https://www.oldbaileyonline.org/browse.jsp?div=t16921207-16 (accessed 9 December 2015).

41 e.g. Fay Bound [Alberti], 'An "Uncivill" Culture: Marital Violence and Domestic Politics in York, *c.*1660–*c.*1760', in *Eighteenth-Century York: Culture, Space and Society*, ed. Mark Hallett and Jane Rendall (York: Borthwick, 2003), 50–8.

42 https://www.oldbaileyonline.org/print.jsp?div=t17180910-77 (accessed 1 December 2015).

43 Donald G. Dutton, *The Domestic Assault of Women: Psychological and Criminal Justice Perspectives* (Vancouver: UBC Press, 1995), 89.

44 See Lisa Gormley, 'Eradicating Stereotypes: Narrowing the Space between the Ideals and Reality for Women's Equality', *Journal of Human Rights Practice*, 3 (2011), 220–8 and Martha Albertson Fineman, *The Public Nature of Private Violence: The Discovery of Domestic Abuse* (New York: Routledge, 1994).

45 William Averell, *Mervailous Combat of Contrarieties* (1538) cited in Jonathan Gil Harris, *Foreign Bodies and the Body Politic: Discourses of Social Pathology in Early Modern England* (Cambridge: Cambridge University Press, 2006), 43.

46 Archbishop William Laud, cited in David Cressy, *Dangerous Talk: Scandalous, Seditious and Treasonable Speech in Pre-Modern England* (Oxford: Oxford University Press, 2010), 10.

47 Cressy, *Dangerous Talk*, ch. 1 and David Cressy, 'Demotic Voices and Popular Complaint in Elizabethan and Early Stuart England', *Journal of Early Modern Studies*, 2 (2013), 47–62, at 53.

48 Laura Gowing, *Gender Relations in Early Modern England* (Harlow, Essex: Pearson, 2012), 75.

49 See the discussion in Cressy, *Dangerous Talk*, 10.

50 Gowing, *Domestic Dangers* and Bernard Stuart Capp, *When Gossips Meet: Women, Family and Neighbourhood in Early Modern England* (Oxford: Oxford University Press, 2003).

51 Tara MacDonald, '"She'd Give her Two Ears to Know": The Gossip Economy in Ellen Wood's *St Martin's Eve*', in *Economic Women: Essays on Desire and Dispossession in Nineteenth-Century British Culture*, ed. Jill Rappoport and Lana L. Dalley (Colombus, OH: Ohio State University Press, 2013), 179–92.

52 For a recent discussion of the work of Fougasse, see Jo Fox, 'Careless Talk: Tensions within British Domestic Propaganda during the Second World War', *The Journal of British Studies*, 51 (2012), 936–66.

53 Antonia C. Lant, *Blackout: Reinventing Women for Wartime British Cinema* (Princeton, NJ: Princeton University Press, 1991), 76–7.

54 Jean Anthelme Brillat-Savarin, *The Physiology of Taste, or, Meditations on Transcendental Gastronomy* (London: Peter Davies, 1925; repr. New York: Dover Publications, 1960), 21. [French orig., *Physiologie du goût, ou, Méditations de gastronomie transcendante*... (Paris: Sautelet, 1825).]

55 On the French influence on English food culture see Mark M. Smith, *Sensing the Past: Seeing, Hearing, Smelling, Tasting and Touching in History* (Berkeley, CA: University of California Press, 2007), 80–1.

56 Allison K. Deutermann, '"Caviare to the General"? Taste, Hearing, and Genre in *Hamlet*', *Shakespeare Quarterly*, 62 (2011), 230–55, at 238.
57 Maxine Berg and Elizabeth Eger (eds), *Luxury in the Eighteenth Century: Debates, Desires and Delectable Goods* (Basingstoke: Palgrave Macmillan, 2003); Robert W. Jones, *Gender and the Formation of Taste in Eighteenth-Century Britain: The Analysis of Beauty* (Cambridge: Cambridge University Press, 1998).
58 Edmund Burke, [An Essay on the] *Sublime and Beautiful: With an Introductory Discourse Concerning Taste and Other Additions* (Oxford, 1796) and Immanuel Kant, *Critique of Judgement*, trans. J. H. Bernard (Mineola, NY: Dover Publications 2005).
59 Denise Gigante, 'Romanticism and Taste', *Literature Compass*, 4 (2007), 407–19, at 413.
60 Cited in Gigante, 'Romanticism and Taste', 408. See also Denise Gigante (ed.), *Gusto: Essential Writings in Nineteenth-Century Gastronomy* (New York and London: Routledge, 2005).
61 Smith, *Sensing the Past*, 82.
62 Emma Spary, *Eating the Enlightenment: Food and the Sciences in Paris, 1670–1760* (Chicago, IL: University of Chicago Press, 2012), 12–18.
63 Abbé Ange Denis Macquin, *Tabella Cibaria. The Bill of Fare: a Latin poem* (London, 1820), 15.
64 Smith, *Sensing the Past*, 83.
65 Linda M. Bartoshuk, 'History of Taste Research', in *Handbook of Perception*, Vol. 6A: *Tasting and Smelling*, ed. Edward C. Carterette and Morton P. Friedman (New York and London: Academic Press, 1978), ch. 1.
66 Bartoshuk, 'History of Taste Research', 4.
67 Bartoshuk, 'History of Taste Research', 5.
68 Albrecht von Haller, *First Lines of Physiology* (Edinburgh: Charles Elliot, 1786; repr. with introduction by Lester S. King, New York and London: Johnson, 1966), 264. [Latin orig., *Elementa physiologiae corporis humani* (Lausanne: Bousquet et Sociorum, 1757–66).]
69 Haller, *First Lines of Physiology*, 258.
70 M. [Claude-Nicolas] Le Cat, *A Physical Essay on the Senses* (London: R. Griffiths, 1750), 17, 20.
71 Le Cat, *A Physical Essay on the Senses*, 15.
72 Gabriel Gustav Valentin, *A Textbook of Physiology* (London: Henry Renshaw, 1853), discussed in Bartoshuk, 'History of Taste Research', 7.
73 Finger, *Origins of Neuroscience*, 169.
74 L. E. Shore, 'A Contribution to our Knowledge of Taste Sensations', *The Journal of Physiology*, 13 (1892), 191–217.
75 D. P. Hänig, 'Zur Psychophysik des Geschmacksinnes', *Philosophische Studien*, 17 (1901), 576–623.

76 Edwin G. Boring, *Sensation and Perception in the History of Experimental Psychology* (New York New York: Appleton-Century-Crofts, 1942; repr.: Irvington Publishers, 1970).
77 See Christopher U. Smith, *The Biology of Sensory Systems* (Chichester and New York: John Wiley, 2000; 2nd edn, Hoboken, NJ: Wiley, 2008), 241.
78 Bernd Lindemann, Yoko Ogiwara and Yuzo Ninomiya, 'The Discovery of Umami', *Chemical Senses*, 27 (2002), 843–4.
79 Claudio Alexandre Gobatto et al., 'The Monosodium Glutamate (MSG) Obese Rat as a Model for the Study of Exercise in Obesity', *Research Communications in Molecular Pathology and Pharmacology*, 111 (2002), 89–101.
80 Pierre Bourdieu, *Distinction: A Social Critique of the Judgement of Taste*, trans. Richard Nice (1979; London: Routledge, 1984).
81 Bourdieu, *Distinction*, 185, discussed in Alan Warde, *Consumption, Food and Taste* (London: Sage, 1997), 240.
82 See Matthew Kieran, 'The Vice of Snobbery: Aesthetic Knowledge, Justification and Virtue in Art Appreciation', *The Philosophical Quarterly*, 60 (2010), 243–63, and Adam Drewnowski, 'Obesity and the Food Environment: Dietary Energy Density and Diet Costs', *American Journal of Preventive Medicine*, 27 (2004), 154–62.
83 Adam Drewnowski, 'Taste Preferences and Food Intake', *Annual Review of Nutrition*, 17 (1997), 237–53.
84 Michael Moss, *Salt, Sugar, Fat: How the Food Giants Hooked Us* (New York: Allen, 2014).
85 e.g. Cheryl A. Frye et al., 'Menstrual Cycle and Dietary Restraint Influence Taste Preferences in Young Women', *Physiology and Behavior*, 55, (1994), 561–7.
86 Erasmus cited in Ying-chiao Lin, 'The Pathogenic Female Tongue: A Galenic and Paracelsian Diagnosis of *Macbeth*', 臺大文史哲學報, 78 (2013), 209–36, at 214.
87 Cited in Carla Mazzio, 'Sins of the Tongue', 65.
88 Giorgio Baglivi, *The Practice of Physick: Reduc'd to the Ancient Way of Observations* (London: Bell, 1704), 28.
89 G. van Swieten, *Commentaria in Harmanni Boerhaave Aphorismes de Cognoscendis et Curandis Morbus*, 5 vols (Leiden, 1742–72), trans. as *Commentaries on Boerhaave's Aphorisms Concerning the Knowledge and Cure of Diseases*, 17 vols (Edinburgh: Elliot, 1776), III, 20.
90 Cited in Wayne J. Urban, *Gender, Race and the National Education Association: Professionalism and its Limitations* (New York and London: Routledge Falmer, 2000), 807.
91 Jacalyn Duffin, *To See with a Better Eye: A Life of R. T. H. Laennec* (Princeton, NJ: Princeton University, 1998), 238.
92 Cited in Urban, *Gender, Race and the National Education Association*, 807.
93 Casebooks of Peter Mere Latham 1838–71, held at Royal College of Physicians, reference code GB 0113 MS-LATHP. Bound Alberti, *Matters of the Heart*, ch. 5.

94 Bound Alberti, *Matters of the Heart*, 113.

95 See Nancy Holroyde-Downing, 'Mysteries of the Tongue', *Asian Medicine*, 1 (2005), 432–61, and Minah Kim et al., 'Traditional Chinese Medicine Tongue Inspection: An Examination of the Inter- and Intrapractitioner Reliability for Specific Tongue Characteristics', *Journal of Alternative and Complementary Medicine*, 14 (2008), 527–36.

96 Cited in Martin Edwards, 'Put Out Your Tongue! The Role of Clinical Insight in the Study of the History of Medicine', *Medical History*, 55 (2011), 301–6.

97 Walter Kacera, *Ayurvedic Tongue Diagnosis* (Twin Lakes, WI: Lotus Press, 2006), 192.

98 Michael Worboys, *Spreading Germs: Disease Theories and Medical Practice in Britain, 1865–1900* (Cambridge: Cambridge University Press, 2000).

99 T. Gemousakakis et al., 'MEG Evaluation of Taste by Gender Difference', *Journal of Integrative Neuroscience*, 10 (2011), 537–45.

Chapter 8. Fat. So? Gut Knowledge and the Meanings of Obesity

1 Cited in Jan Bondeson, *The Two-Headed Boy and Other Medical Miracles* (Ithaca, NY: Cornell University Press: 2004), 250; the title of this chapter is inspired by Marilyn Wann, *Fat! So? Because you Don't Have to Apologize for your Size!* (Berkeley, CA: Ten Speed Press, 1998) and http://fatso.com (accessed 7 September 2014).

2 Bondeson, *The Two-Headed Boy*, 237, and C. Haynes, *The Life of that Wonderful and Extraordinary Human Mammoth, Daniel Lambert* (London: Guardian, 1883).

3 *The Times*, 2 April 1806, p. 1, col. B.

4 www.europeana.eu/portal/record/9200175/BibliographicResource_3000004688748.html (accessed 1 September 2014).

5 See the discussion in Paul Youngquist, *Monstrosities: Bodies and British Romanticism* (Minneapolis, MN: University of Minnesota Press, 2003), 39–41.

6 Cited in Bondeson, *The Two-Headed Boy*, 243.

7 *The Gentleman's Magazine*, 106 (1809), 681–3.

8 Cited in Bondeson, *The Two-Headed Boy*, 251.

9 http://www.guinnessworldrecords.com/records-3000/heaviest-man (accessed 1 September 2014).

10 David B. Allison et al., 'The Measurement of Attitudes toward and Beliefs about Obese Persons', *International Journal of Eating Disorders*, 10 (2006), 599–607.

11 See Jan Wright, 'BioPower, BioPedagogies and the Obesity Epidemic', in *Biopolitics and the 'Obesity Epidemic': Governing Bodies*, ed. Jan Wright and Valerie Harwood (New York: Routledge, 2009), 1–14, at 1.

12 http://www.who.int/nutrition/publications/obesity/WHO_TRS_894/en/ (accessed 1 June 2014).

13 Sander Gilman, *Fat: A Cultural History of Obesity* (Hoboken, NJ: Wiley, 2013), 15.
14 http://www.nichd.nih.gov/health/topics/obesity/conditioninfo/Pages/risk.aspx#f1 (accessed 7 March 2014).
15 http://www.nhs.uk/news/2013/02February/Pages/Latest-obesity-stats-for-England-are-alarming-reading.aspx (accessed 7 March 2014).
16 Mervyn Deitel, 'From the Editor's Desk: The International Obesity Task Force and "Globesity" ', *Journal of Obesity Surgery*, 12 (2002), 613–14.
17 Timi B. Gustafson and David B. Sarwer, 'Childhood Sexual Abuse and Obesity', *Obesity Reviews*, 5 (2004), 129–35.
18 Susie Orbach, *Fat is a Feminist Issue: The Anti-Diet Guide to Permanent Weight Loss* (New York: Paddington Press, 1978).
19 Denise T. D. Ridder et al., 'Taking Stock of Self-Control: A Meta-Analysis of how Trait Self-Control Relates to a Wide Range of Behaviors', *Personality and Social Psychology Review*, 16 (2012), 76–99.
20 See Gerbrand C. M. van Hout et al., 'Psychological Profile of the Morbidly Obese', *Obesity Surgery*, 14 (2004), 579–88.
21 Georges Vigarello, *The Metamorphoses of Fat: A History of Obesity*, trans. C. Jon Delogu (New York: Columbia University Press, 2013), 3.
22 Joan P. Alcock, 'Gluttony: The Fifth Deadly Sin', in *The Fat of the Land: Proceedings of the Oxford Symposium on Food and Cookery 2002*, ed. Harlan Walker (Bristol: Footwork, 2003), 11–21.
23 Tobias Venner, *Via recta ad vitam longam, or, A Plaine Philosophical Discourse* (London: Griffin, 1620).
24 Venner, *Via recta ad vitam longam*, 267–78.
25 Chris Bain, 'Commentary: What's Past is Prologue', *International Journal of Epidemiology*, 35 (2006), 16–17.
26 Hari Sharma and Hari M. Chandola, 'Obesity in Ayurveda: Dietary, Lifestyle and Herbal Considerations', in *Bioactive Food as Dietary Interventions for Diabetes*, ed. Ronald Ross Watson and Victor R. Preedy (Boston, MA: Elsevier, 2013), 463–80.
27 Niki S. Papavramidou et al., 'Galen on Obesity: Etiology, Effects, and Treatment', *World Journal of Surgery*, 28 (2004), 631–5.
28 Venner, *Via recta ad vitam longam*, 268–9.
29 Papavramidou et al., 'Galen on Obesity', 632.
30 Papavramidou et al., 'Galen on Obesity', 634.
31 Malcolm Flemyng, *A Discourse on the Nature, Causes, and Cure of Corpulency* (London: Davis and Reymers, 1760).
32 Flemyng, *A Discourse*, 10–13.
33 Flemyng, *A Discourse*, 7.
34 Anita Guerrini, *Obesity and Depression in the Enlightenment: The Life and Times of George Cheyne* (Norman, OK: University of Oklahoma Press, 2000).

35 George Cheyne, *An Essay on Health and Long Life* (London: G. Strahan and J. Leake, 1724; repr. New York: Edward Gillespy, 1813).
36 George Cheyne, *The Natural Method of Cureing the Diseases of the Body, and the Disorders of the Mind Depending on the Body*, 3rd edn (London: Strahan, 1742).
37 George Cheyne, *The English Malady, or, A Treatise of Nervous Diseases of All Kinds, as Spleen, Vapours, Lowness of Spirits, Hypochondriacal and Hysterical Distempers* (London: Strahan, 1733).
38 Cheyne, *English Malady*, 102.
39 Cheyne, *English Malady*, a2–a3.
40 Anita Guerrini, 'The Hungry Soul: George Cheyne and the Construction of Femininity', *Eighteenth-Century Studies*, 32 (1999), 279–91.
41 Vigarello, *The Metamorphoses of Fat*, 104.
42 Sidney Mintz, *Sweetness and Power: The Place of Sugar in Modern History* (London: Penguin, 1986), 6.
43 Carolyn Thomas de la Peña, *The Body Electric: How Strange Machines Built the Modern American* (New York: New York University Press, 2003), esp. 23.
44 Deborah Lupton, *Medicine as Culture: Illness, Disease and the Body in Western Societies* (Thousand Oaks, CA: Pine Forge, 2012), 60.
45 Kenneth L. Caneva, *Robert Mayer and the Conservation of Energy* (Princeton, NJ: Princeton University Press, 1993); James L. Hargrove, 'History of the Calorie in Nutrition', *The Journal of Nutrition*, 136 (2006), 2957–61.
46 Ancel Keys et al., 'Indices of Relative Weight and Obesity', *Journal of Chronic Diseases*, 25 (1972), 329–43; Margaret Ashwell, 'Charts Based on Body Mass Index and Waist-to-Height Ratio to Assess the Health Risks of Obesity: A Review', *Open Obesity Journal*, 3 (2011), 78–84.
47 Ian Janssen et al., 'Waist Circumference and not Body Mass Index Explains Obesity-Related Health Risk', *The American Journal of Clinical Nutrition*, 79 (2004), 379–84.
48 Bound Alberti, *Matters of the Heart*, ch. 2.
49 Garabed Eknoyan, 'Adolphe Quetelet (1796–1874): The Average Man and Indices of Obesity', *Nephrology Dialysis Transplantation*, 23 (2008), 47–51.
50 Geoffrey Cannon, 'The Rise and Fall of Dietetics and of Nutrition Science, 4000 BCE–2000 CE', *Public Health Nutrition*, 8 (2005), 701–5.
51 William Harvey, *On Corpulence in Relation to Disease: With Some Remarks on Diet* (London: Renshaw, 1872), vi.
52 William Banting, *Letter on Corpulence* (London: Harrison, 1864), 5.
53 Francis Edmund Anstie, 'Corpulence', *Cornhill Magazine*, 7 (1863), 457–68.
54 Anstie, 'Corpulence', 457–8.
55 Anstie, 'Corpulence', 458–9.
56 Anstie, 'Corpulence', 459–60.
57 Banting, *Letter on Corpulence*, 19.

58 Banting, *Letter on Corpulence*, 24.
59 Michelle Mouton, '"Doing Banting": High-Protein Diets in the Victorian Period and Now', *Studies in Popular Culture*, 24 (2001), 17–32.
60 See Roy F. Baumeister and John Tierney, *Willpower: Rediscovering our Greatest Strength* (London: Allen Lane, 2012), introduction.
61 Samuel Smiles, *Self-Help: With Illustrations of Character and Conduct* (London: 1859), 94.
62 For instance, Frank Channing Haddock, *Power of Will: A Practical Companion Book for Unfoldment of the Powers of Mind*, rev. edn (Meriden: Pelton, 1919).
63 Susan Bordo, 'Beyond the Anorexic Paradigm: Re-Thinking "Eating" Disorders', in *Routledge Handbook of Body Studies*, ed. Bryan S. Turner (London: Routledge, 2014), 244–63.
64 Lindsey A. Guerrero, 'The Force-Feeding of Young Girls: Mauritania's Failure to Enforce Preventative Measures and Comply with the Convention on the Elimination of All Forms of Discrimination against Women', *Transnational Law and Contemporary Problems*, 21 (2012), 879–910.
65 Andrew J. Hill, 'Prevalence and Demographics of Dieting', in *Eating Disorders and Obesity: A Comprehensive Handbook*, ed. Christopher G. Fairburn and Kelly D. Brownell (New York and London: Guildford Press, 2002), ch. 14, esp. p. 81.
66 For a critique see Kelly D. Brownell, 'Whether Obesity Should be Treated', *Health Psychology*, 12 (1993), 339–41.
67 For an introduction to the politics behind food production see J. Eric Oliver, *Fat Politics: The Real Story Behind America's Obesity Epidemic* (Oxford: Oxford University Press, 2006).
68 Peter Stearns, *Fat History: Bodies and Beauty in the Modern West* (New York: New York University Press: 2002).
69 Amy Erdman Farrell, *Fat Shame: Stigma and the Fat Body in American Culture* (New York: New York University Press: 2011), 40.
70 http://www.slowfood.org.uk (accessed 22 March 2015).
71 http://www.imt.ie/news/uncategorized/2007/08/its-time-to-stigmatise-fat-people.html (accessed 1 June 2014); Peter Hopkins, 'Everyday Politics of Fat', *Antipode*, 44 (2012), 1227–46.
72 Daniel Black, *Embodiment and Mechanisation: Reciprocal Understandings of Body and Machine from the Renaissance to the Present* (Farnham, Surrey: Ashgate, 2014); Fay Bound Alberti, 'Bodies, Hearts, and Minds: Why Emotions Matter to Historians of Science and Medicine', *Isis*, 100 (2009), 798–810.
73 Marion Nestle and Michael F. Jacobson, 'Halting The Obesity Epidemic: A Public Health Policy Approach', *Public Health Reports*, 115 (2000), 12; Colin Hector, 'Nudging towards Nutrition: Soft Paternalism and Obesity-Related Reform', *Food and Drug Law Journal*, 67 (2012), 103.

74 Brian Wansink and Jeffery Sobal, 'Mindless Eating: The 200 Daily Food Decisions We Overlook', *Environment And Behavior*, 39 (2007), 106–23.
75 Robert H. Lustig, 'Childhood Obesity: Behavioral Aberration or Biochemical Drive? Reinterpreting the First Law of Thermodynamics', *Nature Reviews Endocrinology*, 2 (2006), 447–58; George A. Bray and Catherine M. Champagne, 'Beyond Energy Balance: There is More to Obesity than Kilocalories', *Journal of the American Dietetic Association*, 105 (2005), 17–23.
76 Bordo, 'Beyond the Anorexic Paradigm', 252.
77 Geoff Watts, 'Obesity: In Search of Fat Profits', *BMJ*, 334 (2007), 1298–9.
78 Robert H. Lustig et al., 'Public Health: The Toxic Truth about Sugar', *Nature*, 482 (2012), 27–9.
79 Lustig et al., 'Public Health', 28.
80 Richard J. Johnson et al. 'Potential Role of Sugar (Fructose) in the Epidemic of Hypertension, Obesity and the Metabolic Syndrome, Diabetes, Kidney Disease, and Cardiovascular Disease', *The American Journal of Clinical Nutrition*, 86 (2007), 899–906.
81 Arya M. Sharma and Raj Padwal, 'Obesity is a Sign—Overeating is a Symptom: An Aetiological Framework for the Assessment and Management of Obesity', *Obesity Reviews*, 11 (2010), 362–70.
82 Stefan R. Borntein, M. L. Wong and Julio Licinio, 'Perspective: 150 years of Sigmund Freud: What would Freud have said about the Obesity Epidemic?' *Molecular Psychiatry*, 11 (2006), 1070–2.
83 Vincent L. Wester et al., 'Long-Term Cortisol Levels Measured in Scalp Hair of Obese Patients', *Obesity*, 22 (2014), 1956–8.
84 Willis F. Overton and David Stuart Palermo (eds), *The Nature and Ontogenesis of Meaning* (Hillsdale, NJ: Erlbaum, 1994), 159. For an introduction to the historical varieties of intuition, see Lisa M. Osbeck and Barbara S. Held (eds), *Rational Intuition* (Cambridge: Cambridge University Press, 2014).
85 R. James Hankinson, 'Galen's Anatomy of the Soul', *Phronesis*, 36 (1991), 197–233.
86 David Ricky Matsumoto, *Unmasking Japan: Myths and Realities About the Emotions of the Japanese* (Stanford, CA: Stanford University Press, 1996), 127.
87 *OED*, 'Guttes'.
88 Robert W. Dent, *Shakespeare's Proverbial Language: An Index* (Berkeley, CA: University of California Press, 1981), p 392.
89 *Henry V*. IV. iii. 34–5.
90 Carol Levin, '*The Heart and Stomach of a King*': *Elizabeth I and the Politics of Sex and Power* (Philadelphia, PA: University of Philadelphia Press, 1994), 1.
91 See Jan Purnis, The Stomach and Early Modern Emotion', *University of Toronto Quarterly*, 79 (2010), 800–18.
92 *The Times*, 22 July 1969, p. 2.

93 Galen, *On Diseases and Symptoms*, ed. Ian Johnson (Cambridge: Cambridge University Press, 2006), 276.
94 Cited in Eve Keller, *Generating Bodies and Gendered Selves: The Rhetoric of Reproduction in Early Modern England* (Seattle, WA: University of Washington Press, 2007), 61.
95 Purnis, The Stomach and Early Modern Emotion', 809.
96 Andrzej Śródka, 'The Short History of Gastroenterology', *Journal of Physiology and Pharmacology*, 53 (2003), 9–21.
97 Robert Burton, *The Anatomy of Melancholy* (London: 1621), 'Argument of the Frontispiece', x.
98 http://www.ncbi.nlm.nih.gov/pubmedhealth/PMH0041922 (accessed 31 October 2014).
99 Emma C. Spary, *Eating the Enlightenment: Food and the Sciences in Paris, 1670–1760* (Chicago, IL: University of Chicago Press, 2012), ch. 1; Emma C. Spary, *Feeding France: New Sciences of Food, 1760–1815* (Cambridge: Cambridge University Press, 2014), 92.
100 Walter Pagel, *Joan Baptista van Helmont: Reformer of Science and Medicine* (Cambridge: Cambridge University Press, 2002), 129.
101 Charles E. Dinsmore, 'Lazzaro Spallanzani: Concepts of Generation and Regeneration', in *A History of Regeneration Research: Milestones in the Evolution of a Science*, ed. Charles E. Dinsmore (Cambridge: Cambridge University Press, 2007), 67–90.
102 Ian Miller, *A Modern History of the Stomach: Gastric Illness, Medicine and British Society 1800–1950* (London: Pickering and Chatto, 2011).
103 William Beaumont *Experiments and Observations on the Gastric Juice, and the Physiology of Digestion* (Edinburgh: Maclachlan & Stewart, 1838).
104 Beaumont, *Experiments and Observations*, 139.
105 James C. Whorton, *Inner Hygiene: Constipation and the Pursuit of Health in Modern Society* (Oxford University Press: 2000), xi–xii.
106 *Dublin Medical Journal*, 13 (1838), 334–5.
107 Walter B. Cannon, 'The Influence of Emotional States on the Functions of the Alimentary Canal', *The American Journal of the Medical Sciences*, 137 (1909), 480–6, at 482; Lukas Van Oudenhove and Qasim Aziz, 'The Role of Psychosocial Factors and Psychiatric Disorders in Functional Dyspepsia', *Nature Reviews Gastroenterology and Hepatology*, 10 (2013), 158–67.
108 Paul Kline, *Psychology and Freudian Theory: An Introduction* (London: Routledge, 2014), 13.
109 Bound Alberti, *Matters of the Heart*, ch. 2.
110 *OED*, 'Gut-check'.
111 Malcolm Gladwell, *Blink: The Power of Thinking Without Thinking* (London: Penguin, 2006); Gerd Gigerenzer, *Gut Feelings: The Intelligence of the Unconscious*

(London: Penguin, 2007); Daniel Kahneman, *Thinking Fast and Slow* (London: Penguin, 2012).

112 Louann Brizendine, *The Female Brain* (London: Random House, 2007), 161.

113 Emeran A. Mayer, 'Gut Feelings: The Emerging Biology Of Gut–Brain Communication', *Nature Reviews Neuroscience*, 12 (2011), 453–66.

114 Michael D. Gershon, *The Second Brain* (New York: Harper Collins, 1998).

115 Mayer, 'Gut Feelings'.

116 Robinson, Byron, *The Abdominal and Pelvic Brain* (Hammond, IN: Betz, 1907).

117 Robinson, *Abdominal and Pelvic Brain*, 123–6.

118 John Newport Langley, 'The Autonomic Nervous System', *Brain*, 26 (1903), 1–26; David L. McMillin et al., 'The Abdominal Brain and Enteric Nervous System', *The Journal of Alternative and Complementary Medicine*, 5 (1999), 575–86.

119 Mayer, 'Gut Feelings'.

120 Ciaran D. Corcoran et al., 'Vagus Nerve Stimulation in Chronic Treatment-Resistant Depression: Preliminary Findings of an Open-Label Study', *The British Journal of Psychiatry*, 189 (2006), 282–3.

121 Van Oudenhove and Aziz, 'The Role of Psychosocial Factors'.

122 Almudena Sánchez-Villegas et al., 'Dietary Fat Intake and the Risk of Depression: The SUN Project', *PLoS ONE*, 6 (2011), e16268.

123 John F. Cryan, and Siobhain M. O'Mahony, 'The Microbiome-Gut-Brain Axis: From Bowel to Behaviour', *Neurogastroenterology and Motility*, 23 (2011), 187–92.

124 Jane A. Foster and Karen-Anne McVey Neufeld, 'Gut–Brain Axis: How the Microbiome Influences Anxiety and Depression', *Trends in Neurosciences*, 36 (2013), 305–12.

125 Syed S. Hussain and Stephen R. Bloom, 'The Regulation of Food Intake by the Gut–Brain Axis: Implications for Obesity', *International Journal of Obesity*, 37 (2013), 625–33.

126 Neil B. Ruderman et al,. 'AMPK, Insulin Resistance, and the Metabolic Syndrome', *The Journal of Clinical Investigation*, 123 (2013), 2764–72.

Conclusion. Towards Embodiment

1 Nick Crossley, 'Phenomenology and the Body,' in *Routledge Handbook of Body Studies*, ed. Bryan S. Turner (London and New York: Routledge, 2014), 130–143, at 131.

2 Antonio R. Damasio, *Descartes' Error: Emotion, Reason and the Human Brain* (London: Picador, 1994).

3 Emeran A. Mayer, 'Gut Feelings: The Emerging Biology of Gut-Brain Communication', *Nature Reviews Neuroscience*, 12 (2011), 453–466.

4 Françoise Baylis, 'A Face is Not Just Like a Hand: Pace Barker', *The American Journal of Bioethics*, 4 (2004), 30–32. I am currently working on a cultural history of face and hand transplants.
5 Jasmijn M. Herruer et al., 'Negative Predictors for Satisfaction in Patients Seeking Facial Cosmetic Surgery: A Systematic Review', *Plastic and Reconstructive* Surgery, 135 (2015), 1596–1605.
6 Ian Burkitt, *Bodies of Thought: Embodiment, Identity and Modernity* (London: Sage, 1999), 2.
7 Crossley, 'Phenomenology and the Body', 136.
8 Ian Burkitt, *Emotions and Social Relations* (London: Sage, 2014), 13.
9 For an introduction see: Ludmilla Jordanova, 'Medicine and the Visual Arts: Overview' in *Medicine, Health and the Arts: Approaches to the Medical Humanities*, ed. Victoria Bates, Alan Bleakley, and Sam Goodman (Abingdon, Oxon: Routledge, 2014), 41–63.
10 Susan Sontag, *Illness as Metaphor and AIDS and its Metaphors* (New York: Double Day 1990).
11 Sean Nee, 'The Great Chain of Being', *Nature*, 435 (2005), 429.
12 William Harvey, *Exercitatio anatomica de motu cordis et sanguinis in animalibus* (Frankfurt am Main: Sumptibus Guilielmi Fitzeri, 1628), 3.
13 Fay Bound Alberti, 'Introduction: Emotions in the Early Modern Medical Tradition', in *Medicine, Emotion and Disease, 1700–1950*, ed. Fay Bound Alberti (Basingstoke: Palgrave Macmillan, 2006), xiii–xxviii, at xix.
14 Cited in Phiroze Hansotia, 'A Neurologist Looks at Mind and Brain: "The Enchanted Loom" ', *Clinical Medicine & Research*, 1 (2003), 327–32, at 329.
15 Antonio R. Damasio, *The Feeling of What Happens: Body, Emotion and the Making of Consciousness* (London: Vintage, 2000), 11.
16 Burkitt, *Emotions and Social Relations*, 91.
17 Emily Martin, 'Medical Metaphors of Women's Bodies: Menstruation and Menopause', *International Journal of Health Services*, 18 (1988), 237–54.
18 George Lakoff and Mark Johnson, *Metaphors We Live By* (Chicago, IL: University of Chicago Press, 2003), 3.
19 Peter R. Huttenlocher, *Neural Plasticity* (Cambridge, MA: Harvard University Press, 2009).
20 Gary W. Small, et al., 'Your Brain on Google: Patterns of Cerebral Activation During Internet Searching', *American Journal of Geriatric Psychiatry*, 17 (2009), 116–26.
21 Wendy Moyle et al., 'Dementia and Loneliness: An Australian Perspective', *Journal of Clinical Nursing*, 20 (2011), 1445–53.
22 Rees and Weil, 'Integrated Medicine', p. 119.
23 John Preston Wilson, *Trauma, Transformation, and Healing: An Integrated Approach To Theory Research and Post Traumatic Therapy* (London: Routledge, 2014).

24 Dawson Church, 'The Treatment of Combat Trauma in Veterans Using EFT (Emotional Freedom Techniques): A Pilot Protocol', *Traumatology: An International Journal*, 16 (2010), 55–65.

25 Thomas Fuchs, 'The Phenomenology of Body Memory', in *Body Memory, Metaphor and Movement*, edited by Sabine C. Koch et al. (Amsterdam and Philadelphia, PA: Benjamins 2012), 9–22; Eric J. Nestler, 'Cellular Basis of Memory for Addiction', *Dialogues in Clinical Neuroscience*, 15 (2013), 431–43, at 431.

26 Robert Anderson, 'A Case Study in Integrative Medicine: Alternative Theories and the Language of Biomedicine', *The Journal of Alternative and Complementary Medicine*, 5 (1998), 165–73, and Kerryn Phelps, General Practice: The Integrative Approach (Sydney: Elsevier 2011).

27 Vyjeyanthi S. Periyakoil, 'Using Metaphors in Medicine', *Journal of Palliative Medicine* 11 (2008), 842–4.

28 B. Barrett et al., 'Themes of Holism, Empowerment, Access, and Legitimacy Define Complementary, Alternative, and Integrative Medicine in Relation to Conventional Biomedicine', *The Journal of Alternative & Complementary Medicine*, 9 (2003), 937–47 at 940, 941.

29 Luke S. DeHart, 'Improving Public Health from an Epicurean Perspective', *Journal of Public Health* (2014), first published online December 21, 2014, doi:10.1093/pubmed/fdu106.

30 Jeanine J. Stefanucci et al., 'Follow Your Heart: Emotion Adaptively Influences Perception', *Social and Personality Psychology Compass*, 5 (2011), 296–308.

FURTHER READING

Akmal, Mohd, et al., 'Ibn Nafis: A Forgotten Genius in the Discovery of Pulmonary Blood Circulation', *Heart Views*, 11 (2010), 26–30.

Alberti, Samuel J. M. M., *Morbid Curiosities: Medical Museums in Nineteenth-Century Britain* (Oxford and New York: Oxford University Press, 2011).

Aldersey-Williams, Hugh, *Anatomies: The Human Body, its Parts and the Stories They Tell* (London: Penguin, 2012).

Altick, Richard, *The Shows of London* (Cambridge, MA: Harvard University Press, 1978).

Angell, Marcia, 'Evaluating the Health Risks of Breast Implants: The Interplay of Medical Science, the Law, and Public Opinion', *New England Journal of Medicine*, 334 (1996), 1513–18.

Angell, Marcia, *Science on Trial: The Clash of Medical Evidence and the Law in the Breast Implant Case* (London: Norton, 1997).

Angier, Natalie, *Woman: An Intimate Biography* (Boston, MA: Houghton Mifflin, 1999).

Avicenna, *The Canon of Medicine* (New York: AMS Press, 1973).

Bamborough, John Bernard, *The Little World of Man* (London: Longmans, Green, 1952).

Bames, H. O., 'Breast Malformations and a New Approach to the Problem of the Small Breast', *Plastic and Reconstructive Surgery*, 5 (1950), 499–506.

Banks, Joseph, *Journal of the Right Hon. Sir Joseph Banks During Captain Cook's First Voyage in HMS Endeavour in 1768–71*, ed. Joseph D. Hooker (London: Macmillan, 1896).

Banting, William, *Letter on Corpulence* (London: Harrison, 1863).

Bateman, Thomas, *A Practical Synopsis of Cutaneous Diseases: According to the Arrangement of Dr Willan* (London: Longman, Hurst, Rees, Orme, and Brown, 1813).

Bateman, Thomas, *Delineations of Cutaneous Diseases* (London: Longman, 1817).

Beaumont, William, *Experiments and Observations on the Gastric Juice, and the Physiology of Digestion* (Edinburgh: Maclachlan & Stewart, 1838).

Bell, Charles, *Idea of a New Anatomy of the Brain: Submitted for the Observations of his Friends* (London: Strahan and Preston, 1811).

Berker, Ennis A., et al., 'Translation of Broca's 1865 Report: Localization of Speech in the Third Left Frontal Convolution', *Archives of Neurology*, 43 (1986), 1065–72.

Bernasconi, Robert, and Tommy L. Lott (eds), *The Idea of Race* (Indianapolis, IN: Hackett, 2000).

Bertelli, Sergio, *The King's Body: Sacred Rituals of Power in Medieval and Early Modern Europe*, trans. R. Burr Litchfield (University Park, PA: Pennsylvania State University Press, 2001).

Bigelow, Henry Jacob, 'Dr. Harlow's Case of Recovery from the Passage of an Iron Bar through the Head', *American Journal of the Medical Sciences*, 19 (1850), 13–22.

Blay, Yaba Amgborale, 'Skin Bleaching and Global White Supremacy: By Way of Introduction', *Journal of Pan African Studies*, 4 (2011), 4–46.

Bondeson, Jan, *The Two-Headed Boy and Other Medical Miracles* (Ithaca, NY, and London: Cornell University Press: 2004).

Bondurant, Stuart, et al. (eds), *Safety of Silicone Breast Implants* (Washington DC: National Academies Press, 2000).

Bordo, Susan, *Unbearable Weight: Feminism, Western Culture, and the Body* (Berkeley, CA: University of California Press, 2003).

Bordo, Susan, 'Beyond the Anorexic Paradigm: Re-Thinking "Eating" Disorders', in *Routledge Handbook of Body Studies*, ed. Bryan S. Turner (Abingdon-on-Thames: Routledge, 2014), 244–63.

Bound [Alberti], Fay, 'Writing the Self? Love and the Letter in England, *c*.1660–*c*.1760', *Literature and History*, 11 (2002), 1–19.

Bound [Alberti], Fay, 'An Angry and Malicious Mind: Narratives of Slander at the Church Courts of York, *c*.1660–*c*.1760', *History Workshop Journal*, 46 (2003), 59–77.

Bound [Alberti], Fay, 'An "Uncivill" Culture: Marital Violence and Domestic Politics in York, *c*.1660–*c*.1760', in *Eighteenth-Century York: Culture, Space and Society*, ed. Mark Hallett and Jane Rendall (York: Borthwick, 2003), 50–8.

Bound Alberti, Fay, 'The Emotional Heart: Locating the Soul', in *The Heart*, ed. James Peto (New Haven, CT: Yale University Press, 2006), 125–42.

Bound Alberti, Fay, 'Introduction: Emotions in the Early Modern Medical Tradition', in *Medicine, Emotion and Disease, 1700–1950*, ed. Fay Bound Alberti (Basingstoke: Palgrave Macmillan, 2006), xiii–xxviii.

Bound Alberti, Fay, 'Bodies, Emotions and Historians, or, Why the History of the Heart Matters to Historians of Science and Medicine', *Isis*, 100 (2009), 798–810.

Bound Alberti, Fay, *Matters of the Heart: History, Medicine and Emotion* (Oxford: Oxford University Press, 2010).

Bound Alberti, Fay (ed.), *Medicine, Emotion and Disease, 1700–1950* (Basingstoke: Palgrave Macmillan, 2006).

Bourdieu, Pierre, *Distinction: A Social Critique of the Judgement of Taste*, trans. Richard Nice (London: Routledge, 1984).

Bourke, Joanna, *The Story of Pain: From Prayer to Painkillers* (Oxford: Oxford University Press, 2014).

Boyadjian, Noubar, *The Heart: Its History, Its Symbolism, Its Iconography and Its Diseases*, trans. Agnes Hall (Antwerp: Esco Books, 1985).
Bramwell, Byrom, *Diseases of the Heart and Thoracic Aorta* (Edinburgh: Pentland, 1884).
Braun, Virginia, 'The Women Are Doing it for Themselves: The Rhetoric of Choice and Agency around Female Genital "Cosmetic Surgery"', *Australian Feminist Studies*, 24 (2009), 233–49.
Brizendine, Louann, *The Female Brain* (London: Bantam, 2006).
Brumberg, Joan, *Fasting Girls: The History of Anorexia Nervosa* (Cambridge, MA: Harvard University Press, 1988).
Bunn, Geoffrey C. *The Truth Machine: A Social History of the Lie Detector* (Baltimore, MD: Johns Hopkins University Press, 2012).
Bunzel, Benjamin, et al., 'Does Changing the Heart Mean Changing Personality? A Retrospective Inquiry on 47 Heart Transplant Patients', *Quality of Life Research*, 1 (1992), 251–6.
Burgess, Thomas H., *The Physiology or Mechanism of Blushing: Illustrative of the Influence of Mental Emotion on the Capillary Circulation* (London: Churchill, 1839).
Burkitt, Ian, *Social Selves: Theories of the Social Formation of Personality* (London: Sage, 1991).
Burkitt, Ian, *Bodies of Thought: Embodiment, Identity and Modernity* (London: Sage, 1999).
Burkitt, Ian, *Emotions and Social Relations* (London: Sage, 2014).
Burton, Robert, *Anatomy of Melancholy* (Oxford: John Lichfield and James Short, for Henry Cripps, 1621; repr. New York: New York Review of Books, 2001).
Bynum, William, et al., *A Cultural History of the Human Body*, 6 vols (Oxford: Berg, 2010).
Cannon, Walter Bradford, 'The James-Lange Theory of Emotions: A Critical Examination and an Alternative Theory', *American Journal of Psychology*, 39 (1927), 106–34.
Caputi, Jane, 'The Real "Hot Mess": The Sexist Branding of Female Pop Stars', *Sex Roles*, 70 (2014), 439–41.
Charleton, Walter, *Natural History of the Passions* (London: James Magnes, 1674).
Cheyne, George, *The English Malady, or, A Treatise of Nervous Diseases of All Kinds, as Spleen, Vapours, Lowness of Spirits, Hypochondriacal and Hysterical Distempers* (London: Strahan, 1733).
Cheyne, George, *The Natural Method of Cureing the Diseases of the Body, and the Disorders of the Mind Depending on the Body*, 3rd edn (London: Strahan, 1742).
Cheyne, George, *An Essay on Health and Long Life* (New York: Edward Gillespy, 1813).
Chopra, Ananda S., 'Ayurveda', in *Medicine across Cultures: History and Practice of Medicine in Non-Western Cultures*, ed. Helaine Selin (Dordrecht and Boston, MA: Kluwer Academic Publishers, 2003), 75–84.

Clarke, Edwin, and Jacyna L. Stephen, *Nineteenth-Century Origins of Neuroscientific Concepts* (Berkeley, CA, and London: University of California Press, 1987).
Connor, Steven, *The Book of Skin* (London: Reaktion, 2004).
Cooper, Sir Astley Paston, *The Anatomy of the Breast* (London: Longman, Orme, Green, Browne, and Longmans, 1840).
Craig, Maxine Leeds, 'Racialized Bodies', in *Routledge Handbook of Body Studies*, ed. Bryan S. Turner (London: Routledge, 2014), 321–32.
Crooke, Helkiah, *Mikrokosmographia: A Description of the Body of Man* (London: William Jaggard, 1615).
Crozier, Ivan, *A Cultural History of the Body in the Modern Age* (Oxford: Berg 2010).
Damasio, Antonio R., *Descartes' Error: Emotion, Reason and the Human Brain* (London: Picador, 1994).
Darwin, Charles, *The Expression of the Emotions in Man and Animals*, 3rd edn, ed. Paul Ekman (London: Fontana, 1999).
Davis, Kathy, *Reshaping the Female Body: The Dilemma of Cosmetic Surgery* (London: Routledge, 1995).
Delpeuch, Francis, *Globesity: A Planet Out of Control?* (London: Routledge, 2013).
Descartes, René, *The Passions of the Soule* (London: trans. and pr. for A. C. and sold by J. Martin and J. Ridley, 1650).
Descartes, Réne, *Discourse on Method and Meditations*, trans. Elizabeth S. Haldane and G. R. T. Ross (Mineola, NY: Dover publications, 2003).
Downame, John, *A Treatise of Anger* (London: William Welby, 1609).
Dror, Otniel E., 'The Scientific Image of Emotion: Experience and Technologies of Inscription', *Configurations*, 7 (1999), 355–401.
Dror, Otniel E., 'Fear and Loathing in the Laboratory and Clinic', in *Medicine, Emotion and Disease, 1700–1950*, ed. Fay Bound Alberti (Basingstoke: Palgrave Macmillan, 2006), 125–43.
Durham, M. Gigi, *The Lolita Effect: The Media Sexualization of Young Girls and What We Can Do About It* (Woodstock, NY: Overlook Press, 2008).
Ellis, Harold, *The Cambridge Illustrated History of Surgery* (Cambridge: Cambridge University Press, 2009).
Ellis, Normandi, *Awakening Osiris: A New Translation of the Egyptian Book of the Dead* (Grand Rapids, MI: Phanes, 1988).
Elwin, Verrier, 'The Vagina Dentata Legend', *British Journal of Medical Psychology*, 19 (1943), 439–53.
Ensler, Eve, *The Vagina Monologues* (London: Random House, 2007).
Fabian, Ann, *The Skull Collectors: Race, Science and America's Unburied Dead* (Chicago, IL: University of Chicago Press, 2010).
Finger, Stanley, *Minds Behind the Brain: A History of the Pioneers and their Discoveries* (Oxford: Oxford University Press, 2000).

Finger, Stanley, *Origins of Neuroscience: A History of Explorations into Brain Function* (Oxford: Oxford University Press, 2001).
Foster, Jane A., and Karen-Anne McVey Neufeld, 'Gut–Brain Axis: How the Microbiome Influences Anxiety and Depression', *Trends in Neurosciences*, 36 (2013), 305–12.
Foucault, Michel, *The Birth of the Clinic: An Archaeology of Medical Perception*, trans. Alan M. Sheridan (London: Tavistock, 1976).
Freud, Sigmund, *On Sexuality: Three Essays on the Theory of Sexuality and Other Works*, trans. James Strachey et al., ed. Angela Richards (London: Penguin-Pelican, 1977; London: Penguin, 1991). [Ger. orig., *Drei Abhandlungen zur Sexualtheorie* (Leipzig and Vienna: Franck Deuticke, 1905).]
Friedland, Roger, 'Looking Through the Bushes: The Disappearance of Pubic Hair', *Huffington Post*, 13 June 2011.
Friedman, Jeff, 'Muscle Memory: Performing Embodied Knowledge', *Routledge Studies in Memory and Narrative*, 10 (2002), 156–80.
Fuchs, Thomas, 'The Phenomenology of Body Memory', in *Body Memory, Metaphor and Movement*, ed. Sabine C. Koch et al. (Amsterdam: Benjamins 2012), 9–22.
Galen, *On the Passions and Errors of the Soul*, trans. Paul W. Harkins (Columbus, OH: Ohio State University Press, 1963).
Galen, *On Diseases and Symptoms*, ed. Ian Johnson (Cambridge: Cambridge University Press, 2006).
Gall, Francis Joseph, *On the Functions of the Brain*, trans. Winslow Lewis, 6 vols (Boston, MA: Marsh, Capen, and Lyon, 1835).
Gallagher, Catherine, and Thomas Walter Laqueur (eds), *The Making of the Modern Body: Sexuality and Society in the Nineteenth Century* (Berkeley, CA: University of California Press, 1986).
Gershon, Michael, *The Second Brain: A Groundbreaking New Understanding of Nervous Disorders of the Stomach and Intestine* (New York: HarperCollins, 1999).
Gigerenzer, Gerd, *Gut Feelings: The Intelligence of the Unconscious* (London: Penguin, 2007).
Gillies, Harold, D., *Plastic Surgery of the Face, Based on Selected Cases of War Injuries of the Face Including Burns* (Oxford: Oxford University Press, 1920).
Gilman, Sander L., *Making the Body Beautiful: A Cultural History of Aesthetic Surgery* (Princeton, NJ: Princeton University Press, 1999).
Gilman, Sander L., *Fat: A Cultural History of Obesity* (London: Polity, 2008).
Gladwell, Malcolm, *Blink: The Power of Thinking Without Thinking* (London: Penguin, 2006).
Glenn, Evelyn Nakano (ed.), *Shades of Difference: Why Skin Color Matters* (Stanford, CA: Stanford University Press, 2009).
Gorham, Deborah, *The Victorian Girl and the Feminine Ideal* (Hoboken, NJ: Taylor and Francis, 2012).

Gowing, Laura, 'Gender and the Language of Insult in Early Modern London', *History Workshop Journal*, 35 (1993), 1–21.
Gowing, Laura, *Gender Relations in Early Modern England* (Harlow: Pearson, 2012).
Grady, Hugh, 'Renewing Modernity: Changing Contexts and Contents of a Nearly Invisible Concept', *Shakespeare Quarterly*, 50 (1999), 268–84.
Greer, Germaine, *The Female Eunuch* (London: MacGibbon and Kee, 1970).
Goude, Jean-Paul, *Jungle Fever*, ed. Harold Hayes (London: Quartet, 1982).
Guerrini, Anita, *Obesity and Depression in the Enlightenment: The Life and Times of George Cheyne* (Norman, OK: University of Oklahoma Press, 2000).
Gylseth, Christopher Hals, and Lars O. Toverud, *Julia Pastrana: The Tragic Story Of The Victorian Ape Woman*, trans. Donald Tumasonis (Stroud: Sutton Publishing, 2004).
Haiken, Elizabeth, *Venus Envy: A History of Cosmetic Surgery* (Baltimore, MD: Johns Hopkins University Press, 1999).
Hall, Mark A., and Carl E. Schneider, 'Patients as Consumers: Courts, Contracts, and the New Medical Marketplace', *Michigan Law Review*, 106 (2008), 643–89.
Haller, Albrecht von, *A Dissertation on the Sensible and Irritable Parts of Animals* (London: Nourse, 1755).
Harrison, Edward, *Pathological and Practical Observations On Spinal Diseases: Illustrated with Cases and Engravings* (London: Underwood, 1827).
Harvey, William, *On Corpulence in Relation to Disease: With Some Remarks on Diet* (London: Renshaw, 1872).
Haskell, Molly, *From Reverence to Rape: The Treatment of Women in the Movies* (Chicago, IL: University of Chicago Press, 1987).
Hayward, Rhodri, *Psychiatry in Modern Britain* (New York: Continuum, 2012).
hooks, bell, *Black Looks: Women, Race and Representation* (Boston, MA: South End Press, 1992).
Hudson, Nicholas, 'From "Nation" to "Race": The Origin of Racial Classification in Eighteenth-Century Thought', *Eighteenth-Century Studies*, 29 (1996), 247–64.
Hudson, Nicholas, '"Hottentots" and the Evolution of European Racism', *Journal of European Studies*, 34 (2004), 308–32.
Huttenlocher, Peter R., *Neural Plasticity* (Cambridge, MA: Harvard University Press, 2009).
Jablonski, Nina G., *Skin: A Natural History* (Berkeley, CA: University of California Press, 2006).
Jacobson, Nora, 'The Socially Constructed Breast: Breast Implants and the Medical Construction of Need', *American Journal of Public Health*, 88 (1998), 1254–61.
James, William, 'What is an Emotion?' *Mind*, 9 (1884), 188–205.
Jasanoff, Sheila, 'Science and the Statistical Victim Modernizing Knowledge in Breast Implant Litigation', *Social Studies of Science*, 32 (2002), 37–69.
Harlow, John M., *Recovery from the Passage of an Iron Bar through the Head* (Boston, MA: Clapp, 1869).

Judowitz, Dalia, *The Culture of the Body: Genealogies of Modernity* (Ann Arbor, MI: University of Michigan Press, 2001).
Keller, Eve, *Generating Body and Gendered Selves: The Rhetoric of Reproduction in Early Modern England* (Seattle, WA: University of Washington Press, 2007).
Keogh, Bruce, *Review of the Regulation of Cosmetic Interventions* (London: Department of Health, 2013).
Knight, Sarah, and Mary Ann Lund, 'Richard Crookback', *Times Literary Supplement*, 6 February 2013, 14–15.
Koedt, Anne, *The Myth of the Vaginal Orgasm* (London: Women's Liberation Movement, 1968).
Kuehn, Manfred, 'Reason and Understanding', in *The Routledge Companion to Eighteenth-Century Philosophy,* ed. Aaron Garrett (London: Routledge, 2014), 167–87, at 175.
Labre, Megdala Peixoto, 'The Brazilian Wax: New Hairlessness Norm for Women?' *Journal of Communication Inquiry*, 26 (2002), 113–32.
Lakoff, George, *Metaphors We Live By* (Chicago, IL: University of Chicago Press, 2003).
Laqueur, Thomas Walter, *Making Sex: Body and Gender from the Greeks to Freud* (Cambridge, MA: Harvard University Press, 1990).
Langley, John Newport, 'The Autonomic Nervous System', *Brain*, 26 (1903), 1–26.
Latham, Melanie, '"If it Ain't Broke, Don't Fix it?" Scandals, "Risk", and Cosmetic Surgery Regulation in the UK and France', *Medical Law Review*, 22 (2014), 384–408.
Lerner, Barron H., 'The Perils of "X-ray Vision": How Radiographic Images Have Historically Influenced Perception', *Perspectives in Biology And Medicine*, 35 (1992), 382–97.
Little, Arthur L., *Shakespeare Jungle Fever: National-Imperial Re-Visions of Race, Rape and Sacrifice* (Stanford, CA: Stanford University Press, 2000).
Logan, Alan C., and Martin Katzman, 'Major Depressive Disorder: Probiotics May be an Adjuvant Therapy', *Medical Hypotheses*, 64 (2005), 533–9.
Lustig, Robert H., et al., 'Public Health: The Toxic Truth About Sugar', *Nature*, 482 (2012), 27–9.
Machida, Masafumi, et al., 'Melatonin: A Possible Role in Pathogenesis of Adolescent Idiopathic Scoliosis', *Spine*, 21 (1996), 1147–52.
Maher, JaneMaree, et al., 'Framing the Mother: Childhood Obesity, Maternal Responsibility and Care', *Journal of Gender Studies,* 19 (2010), 233–47.
Manzoni, Tullio, 'The Cerebral Ventricles, the Animal Spirits and the Dawn of Brain Localization of Function', *Archives italiennes de biologie*, 136 (1998), 103–52.
Martin, Emily, 'Medical Metaphors of Women's Bodies: Menstruation and Menopause', *International Journal of Health Services*, 18 (1988), 237–54.
Martineau, Harriet, *Deerbrook*, 3 vols (London: Edward Moxon, 1839; facs. edn, Virago Press, 1983).
Martyn, Helen, et al., 'Medical Students' Responses to the Dissection of the Heart and Brain: A Dialogue on the Seat of the Soul', *Clinical Anatomy*, 25 (2012), 407–13.

Mayer, Emeran A., 'Gut Feelings: The Emerging Biology of Gut–Brain Communication', *Nature Reviews Neuroscience*, 12 (2011), 453–66.
Mazrui, Ali, 'The Poetics of a Transplanted Heart', *Transition*, 35 (1968), 51–9.
Mazzio, Carla, 'Sins of the Tongue in Early Modern England', *Modern Language Studies*, 28 (1998), 95–124.
McLaughlin, Joseph K., et al., 'The Safety of Silicone Gel-Filled Breast Implants: A Review of the Epidemiologic Evidence', *Annals of Plastic Surgery*, 59 (2007), 569–80.
McMillin, David L., et al., 'The Abdominal Brain and Enteric Nervous System', *The Journal of Alternative and Complementary Medicine*, 5 (1999), 575–86.
Miller, Ian, *A Modern History of the Stomach: Gastric Illness, Medicine and British Society 1800–1950* (London: Pickering and Chatto, 2011).
Miller, Laura, 'Mammary Mania in Japan', *Positions: East Asia Cultures Critique*, 11 (2003), 271–300.
Mintz, Sidney, *Sweetness and Power: The Place of Sugar in Modern History* (London: Penguin, 1986).
Montagu, Ashley, *Touching: The Human Significance of the Skin*, 2nd edn (New York: Harper and Row, 1978).
Moscoso, Javier, *Pain: A Cultural History* (Basingstoke: Palgrave Macmillan, 2012).
Moss, Michael, *Salt, Sugar, Fat: How the Food Giants Hooked Us* (New York: Allen, 2014).
Moyle, Wendy, et al, 'Dementia and Loneliness: An Australian Perspective', *Journal of Clinical Nursing*, 20 (2011), 1445–53.
Muzaffar, Arshad R., and Rod J. Rohrich, 'The Silicone Gel-Filled Breast Implant Controversy: An Update', *Plastic and Reconstructive Surgery*, 109 (2002), 742–8.
Nasuto, Slawomir J., et al., 'Communication as an Emergent Metaphor for Neuronal Operation', in *Computation for Metaphors, Analogy, and Agents*, ed. Chrystopher L. Nehaniv (Berlin: Springer 1999), 365–79.
Nutton, Vivian, *Ancient Medicine* (London and New York: Routledge, 2013).
O'Connor, James P. B., 'Thomas Willis and the Background to *Cerebri Anatome*', *Journal of the Royal Society of Medicine*, 96 (2003), 139–43.
Orbach, Susie, *Fat is a Feminist Issue* (London: Hamlyn, 1982).
Orbach, Susie, *Bodies* (London: Profile Books, 2009).
Panarites, Zoe, 'Breast Implants: Choices Women Thought They Made', *New York School Journal of Human Rights*, 11 (1993), 163–204.
Papavramidou, Niki S., et al., 'Galen on Obesity: Etiology, Effects, and Treatment', *World Journal of Surgery*, 28 (2004), 631–5.
Park, Katherine, 'The Rediscovery of the Clitoris', in *The Body in Parts: Fantasies of Corporeality in Early Modern Europe*, ed. David A. Hillman and Carla Mazzio (Routledge: New York, 1997), 170–93.
Paster, Gail Kern, *Humoring the Body: Emotions and the Shakespearean Stage* (Chicago, IL: University of Chicago Press, 2004).

Pearsall, Paul, et al., 'Changes in Heart Transplant Recipients that Parallel the Personalities of their Donors', *Journal of Near-Death Studies*, 20 (2002), 191–206.

Pitts-Taylor, Victoria, 'Becoming a Cosmetic Surgery Patient: Semantic Instability and the Intersubjective Self', *Studies in Gender and Sexuality*, 10 (2009), 119–28.

Poitevin, Kimberley, 'Inventing Whiteness: Cosmetics, Race and Women in Early Modern England', *Journal for Early Modern Cultural Studies*, 11 (2011), 58–89.

Popkin, Barry M., 'Does Global Obesity Represent a Global Public Health Challenge?' *The American Journal of Clinical Nutrition*, 93 (2011), 232–3.

Powers, Angela, and Julie L. Andsager, 'How Newspapers Framed Breast Implants in the 1990s', *Journalism & Mass Communication Quarterly*, 76 (1999), 551–64.

Purcell, Natalie J., *Violence and the Pornographic Imaginary: The Politics of Sex, Gender and Aggression in Hardcore Pornography* (New York: Routledge, 2012).

Qureshi, Sadiah, 'Displaying Sara Baartman, the "Hottentot Venus"', *History of Science*, 42 (2004), 233–57.

Reed, Edward S., *From Soul to Mind: The Emergence of Psychology from Erasmus Darwin to William James* (New Haven, CT: Yale University Press, 1997).

Rees, Emma L. E., *The Vagina: A Literary and Cultural History* (London: Bloomsbury Academic, 2013).

Rees, Lesley, and Andrew Weil, 'Integrated Medicine: Imbues Orthodox Medicine with the Values of Complementary Medicine', *British Medical Journal*, 322 (2001), 119–20.

Ritner, Robert K., 'The Cult of the Dead', in *Ancient Egypt*, ed. David P. Silverman (Oxford: Oxford University Press, 2003), 132–47.

Rose, F. Clifford, 'Cerebral Localization in Antiquity', *Journal of the History of the Neurosciences*, 18 (2009), 239–47.

Rubin, Miri, *Mother of God: A History of the Virgin Mary* (New York: Allen Lane, 2009).

Rudnytsky, Peter L., 'The "Darke and Vicious Place": The Dread of the Vagina in King Lear', *Modern Philology*, 96 (1999), 291–311.

Russett, Cynthia Eagle, *Sexual Science: The Victorian Construction of Womanhood* (Cambridge, MA, and London: Harvard University Press, 1989).

Scaer, Robert C., *The Body Bears the Burden Trauma, Dissociation and Disease* (New York: Haworth, 2001).

Schiebinger, Londa L., 'Skeletons in the Closet: The First Illustrations of the Female Skeleton in Eighteenth-Century Anatomy', *Representations*, 14 (1986), 42–82.

Schiebinger, Londa, L., *The Mind Has No Sex? Women in the Origins of Modern Science* (Cambridge, MA: Harvard University Press, 1991).

Schiebinger, Londa L., *Nature's Body: Sexual Politics and the Making of Modern Science* (London: Pandora, 1993).

Schiebinger, Londa L., 'Taxonomy for Human Beings', in *The Gendered Cyborg: A Reader*, ed. Gill Kirkup et al. (London and New York: Routledge, 2000), 11–37.

Senderoff, Douglas M., 'Buttock Augmentation with Solid Silicone Implants', *Aesthetic Surgery Journal*, 31 (2011), 320–7.

Shapiro, Jonathan, 'The NHS: The Story So Far (1948–2010)', *Clinical Medicine*, 10 (2010), 335–8.

Sharma, Arya M., and Raj Padwal, 'Obesity is a Sign—Overeating is a Symptom: An Aetiological Framework for the Assessment and Management of Obesity', *Obesity Reviews*, 11 (2010), 362–70.

Shoaib, Britta Ostermeyer, and Bernard M. Patten, 'A Motor Neuron Disease Syndrome in Silicone Breast Implant Recipients', *Journal of Occupational Medicine and Toxicology*, 4 (1995), 155–63.

Shoaib, Britta Ostermeyer, and Bernard M. Patten, 'Human Adjuvant Disease: Presentation as a Multiple Sclerosis-Like Syndrome', *Southern Medical Journal*, 89 (1996), 179–88.

Sikaris, Ken A., 'The Clinical Biochemistry of Obesity', *Clinical Biochemistry Reviews*, 25 (2004), 165–81.

Singer, Charles, 'Galen's Elementary Discourse on Bones', *Proceedings of the Royal Society of Medicine*, 45 (1952), 767–76.

Small, Gary W., et al., 'Your Brain on Google: Patterns of Cerebral Activation during Internet Searching', *The American Journal of Geriatric Psychiatry*, 17 (2009), 116–26.

Sontag, Susan, *Illness as Metaphor and AIDS and its Metaphors* (New York: Double Day 1990).

Spary, Emma C. *Eating the Enlightenment: Food and the Sciences in Paris, 1670–1760* (Chicago, IL: University of Chicago Press, 2012).

Stearns, Peter, *Fat History: Bodies and Beauty in the Modern West* (New York: New York University Press: 2002).

Steinke, Hubert, *Irritating Experiments: Haller's Concept and the European Controversy on Irritability and Sensibility, 1750–90* (Amsterdam: Rodopi, 2005).

Stern, Aglaga, 'Body Piercing: Medical Consequences and Psychological Motivations', *The Lancet*, 361 (2003), 1205–15.

Sylvia, Claire, with William Novak, *A Change of Heart: A Memoir* (London: Little, Brown, 1997).

Tahhan, Diana Adis, 'Blurring the Boundaries between Bodies: Skinship and Bodily Intimacy in Japan', *Japanese Studies*, 30 (2010), 215–30.

Tulloch, Isabel, 'Richard III: A Study in Medical Misrepresentation', *Journal of the Royal Society of Medicine*, 102 (2009), 315–23.

Turner, Bryan S. (ed.), *Routledge Handbook of Body Studies* (Abingdon: Routledge, 2014).

Vanderford, Marsha L., and David H. Smith, *The Silicone Breast Implant Story: Communication and Uncertainty* (Abingdon: Routledge, 2013).

Vesalius, Andreas, *De humani corporis fabrica* [*On the Fabric of the Human Body*] (Basle: Oporini, 1543).

Vidal, Fernando, 'Person and Brain: A Historical Perspective from within the Christian Tradition', *Scripta Varia*, 109 (2007), 3–14.

Wann, Marilyn, *Fat! So? Because you Don't Have to Apologize For Your Size!* (Berkeley, CA: Ten Speed Press, 1998).

Watts, Geoff, 'Obesity: In Search of Fat Profits', *British Medical Journal*, 334 (2007), 1298–9.

Weekes, Debbie, 'Where My Girls At? Black Girls and the Construction of the Sexual', in *All About the Girl: Culture, Power and Identity*, ed. Anita Harris (New York: Routledge, 2004), 141–54.

Weiner, M-F., and J. R. Silver, 'Paralysis as a Result of Traction for the Treatment of Scoliosis: A Forgotten Lesson from History', *Spinal Cord*, 47 (2009), 429–34.

Weiss, Hans-Rudolf, 'Is there a Body of Evidence for the Treatment of Patients with Adolescent Idiopathic Scoliosis (AIS)?' *Scoliosis*, 2 (2007), 19–24.

White, Robert J., et al., 'Cephalic Exchange Transplantation in the Monkey', *Surgery*, 70 (1971), pp. 135–9.

Whitney, Jennifer Dawn, 'Some Assembly Required: Black Barbie and the Fabrication of Nicki Minaj', *Girlhood*, 5 (2012), 141–59.

Willis, Thomas, *The Anatomy of the Brain and Nerves*, ed. William Feindel (Birmingham, AL: Classics of Medicine Library, 1978).

Wilson, John Preston, *Trauma, Transformation, And Healing: An Integrated Approach To Theory Research and Post Traumatic Therapy* (London: Routledge, 2014).

Wolf, Naomi, *The Beauty Myth: How Images of Beauty Are Used against Women* (London: Random House, 2013).

Wright, Jan, 'BioPower, BioPedagogies and the Obesity Epidemic', in *Biopolitics and the 'Obesity Epidemic': Governing Bodies*, ed. Jan Wright and Valerie Harwood (New York: Routledge, 2009), 1–14.

Wright, Thomas, *Passions of the Minde in Generall* (London: VC for WB, 1601; facs. edn, Hildesheim and New York: Olms, 1973).

Wujastyk, Dominic, 'The Science of Medicine', in *The Blackwell Companion to Hinduism*, ed. Gavin Flood (Oxford: Blackwell, 2003), 393–409.

Yalom, Marilyn, *A History of the Breast* (London: Pandora, 1998).

Young, Lola, 'Racializing Femininity', in *Women's Bodies: Discipline and Transgression*, ed. Jane Arthurs and Jean Grimshaw (London: Cassell, 1999), 67–90.

Young, Robert Maxwell, *Mind, Brain, and Adaptation in the Nineteenth Century: Cerebral Localization and its Biological Context from Gall to Ferrier* (Oxford: Oxford University Press, 1970).

Zimmerman, Susan M., *Silicone Survivors: Women's Experiences with Breast Implants* (Philadelphia, PA: Temple University Press, 1998).

PICTURE CREDITS

Figure 1 © University Of Leicester
Figure 2 © The Sun / News Syndication
Figure 3 © The Great Wall of Vagina sculpture by Jamie McCartney 2011
Figure 4 © Wellcome Library, London
Figure 5 © Adrian Brookes Image Wise
Figure 6 Author unknown. Licensed under Public Domain via Commons - https://commons.wikimedia.org/wiki/File:Phineas_Gage_GageMillerPhoto210-02-17_Unretouched_Color_Cropped.jpg#/media/File:Phineas_Gage_GageMillerPhoto210-02-17_Unretouched_Color_Cropped.jpg
Figure 7 © AIP / THE KOBAL COLLECTION
Figure 8 © Wellcome Library, London
Figure 9 © Science Museum, London / Wellcome Images
Figure 10 © BSI-1198105 © JACOPIN / AgeFotostock
Figure 9 © Wellcome Library, London
Figure 12 © Wellcome Library, London

INDEX OF NAMES

SUBJECT INDEX